CHEMISTRY RESEARCH AND APPLICATIONS

PHOTOISOMERIZATION

CAUSES, BEHAVIOR AND EFFECTS

CHEMISTRY RESEARCH AND APPLICATIONS

Additional books and e-books in this series can be found on Nova's website under the Series tab.

CHEMISTRY RESEARCH AND APPLICATIONS

PHOTOISOMERIZATION

CAUSES, BEHAVIOR AND EFFECTS

DIEGO SAMPEDRO
EDITOR

nova
science publishers
New York

Library of Congress Cataloging-in-Publication Data

Names: Sampedro, Diego, editor.
Title: Photoisomerization : causes, behavior, and effects / editor, Diego Sampedro (Universidad de La Rioja, Departamento de Quimica, Madre de Dios, Logrono, Spain).
Description: Hauppauge, New York : Nova Science Publishers, Inc., 2019. |
Series: Chemistry research and applications | Includes bibliographical references and index.
Identifiers: LCCN 2019013911 (print) | LCCN 2019015414 (ebook) | ISBN 9781536153187 () | ISBN 9781536153132 (hardcover)
Subjects: LCSH: Photoisomerization. | Isomerization. | Rearrangements (Chemistry)
Classification: LCC QD471 (ebook) | LCC QD471 .P4975 2019 (print) | DDC 547/.2--dc23
LC record available at https://lccn.loc.gov/2019013911

Published by Nova Science Publishers, Inc. † New York

CONTENTS

PREFACE

Recent years have seen the development of new and exciting applications of light controlled processes in fields as diverse as medicine, material science, energy storage or biochemistry. At the heart of many of these processes underlies a very simple chemical step: a photoisomerization. The isomerization of a chemical double bond is arguably the simplest photochemical process, but still offers many different possibilities to design and control extremely specific applications in complex systems. Many light-activated features in Nature are initiated by this simple chemical change that can trigger a fascinating cascade of signals. For instance, the first event on the process of vision is a very simple C=C photoisomerization. When a photon is absorbed, the light energy is used to induce a chemical change that can be detected by the protein environment that eventually will lead, after a complex series of steps, to a signal received and interpreted by the brain. This is only one sample of a long list of captivating examples of light-controlled processes in Nature mediated by a photoisomerization. Inspired by these amazing transformations, scientists have tried to duplicate and adapt the function of natural systems to perform new applications in different scenarios using again simple photosiomerizations as the key element. As a result, the design and use of molecular devices (especially molecular photoswitches) have attracted much attention in the last two decades. The outcome of this collective effort is a myriad of applications in many different fields that are already having an impact in our everyday life: systems for solar energy storage, control of neuronal processes, light-activated properties in plastics and polymers, regulation of antibiotic or cytotoxic activity of drugs can be all controlled by photoisomerizations.

In this book, the state-of-the-art of the photoisomerization will be presented, together with some fundamental key features to understand, control and tune the photoisomerization of many different chemical structures containing C=C, N=N and C=N double bonds. Experimental and computational tools used to study these processes will be discussed, as the applications of many different compounds in diverse fields.

In Chapter 1, Sampedro and Fernández-Martínez review the synthesis and properties of molecular photoswitches based on E /Z isomerizations. Many different switches have been prepared and used in fascinating applications. From the ubiquitous azobenzenes to the more specific imines, the fundamental key features of several chemical structures used for the construction of molecular devices controlled by a photoisomerization of double bonds are presented.

A complete mechanistic review of the photoisomerization of polypyridyl ruthenium(II) complexes is shown in Chapter 2. Yagi and coworkers present a detailed analysis combining both experimental and computational tools to unveil the key aspects of the isomerization of these compounds with potential application as functional materials.

While cyanine dyes have been known from a long time, their special features and, especially, their applications in lasers, biology and medicine turn them into very relevant species. Tatikolov and Pronkin present in Chapter 3 a detailed overview or the properties and photoisomerization of these molecules.

Azobenzenes have been extensively studied as very relevant molecules capable of photoisomerization. A promising set of compounds, with great potential due to their relevant photochemical properties is the family of azoheteroarenes in which at least one benzene ring in the azobenzene moiety has been changed by an heteroaromatic ring. A detailed review of the properties and applications of these species is presented in Chapter 4 by Venkataramani and coworkers.

Marazzi and García-Iriepa review in Chapter 5 the extensive work performed around retinal photoswitches. Following Nature design of this extremely efficient molecular switch, the detailed mechanistic study performed by computational and experimental techniques has allowed for the preparation of many different compounds with relevant properties. In addition, some relevant applications of these species are presented.

Photoisomerizations have been also used as a key step in organic synthesis. The possibility of controlling by light the geometry of an alkene-containing molecule allows for the design of synthetic routes with a very high selectivity. Oksdath-Mansilla and coworkers present in Chapter 6 a very complete review of the possibilities on this topic.

This book covers from the most fundamental key aspects of the photoisomerization reactions to the applications in different fields. In addition, the use of a wide range of different techniques, from the state-of-the-art computation to spectroscopy and other experimental tools, is thoroughly discussed. Finally, some selected types of photoisomerizing compounds are discussed in depth. Overall, the reader will achieve a comprehensive view of the basic features and the immense potential of these fascinating processes.

Diego Sampedro, PhD

In: Photoisomerization
Editor: Diego Sampedro

ISBN: 978-1-53615-313-2

Chapter 1

MOLECULAR PHOTOSWITCHES BASED ON E/Z ISOMERIZATIONS

Diana Fernández-Martínez and Diego Sampedro*

Department of Chemistry,
Centro de Investigación en Síntesis Química (CISQ),
Universidad de La Rioja, Logroño, Spain

ABSTRACT

The synthesis, study and use of molecular machines in many different applications have seen an extraordinary development in the last two decades. Proof of this is the Nobel Prize in Chemistry awarded in 2016 to Jean-Pierre Sauvage, J. Fraser Stoddart and Bernard L. Feringa *"for the design and synthesis of molecular machines."* Undoubtedly, molecular switches are the most studied and used type of molecular machine and many different systems have been prepared and employed in applications ranging from material science to medicine. Using a very general definition, a molecular switch is a bistable molecule that can be reversibly converted from one state to the other using an external stimulus. In this rapidly growing field, molecular switches are classified attending to two main

---

* Corresponding Author's Email: diego.sampedro@unirioja.es.

criteria: the stimulus that operates the switch and the type of chemical process involved. In this sense, while different stimuli have been used, light has become the most common way to control the activation of these systems, mainly due to the practical advantages that it provides. Regarding the chemical transformations involved, the molecular motions can include relatively simple double-bond isomerizations, but also cyclizations and complex supramolecular interactions.

From the huge number of available species, one specific type of compound has been specially used due to its advantages: photoswitches based on *E*/*Z* isomerizations. In this case, the stimulus that operates the machine is light and the chemical process involved is the photoisomerization of a double bond, usually a C=C or N=N bond. This type of systems can be easily prepared with a high degree of control over most of the relevant properties, from wavelength of absorption to the associated geometrical changes.

Thus, in this contribution we will briefly describe the different types of molecular switches that have been reported with a focus of the systems that imply a photoisomerization. We will review the synthesis of diverse chemical structures together with the different features that can be controlled and tuned. In addition, we will discuss the mechanisms involved in the photoisomerization and how they can be studied and controlled. Also, the use of photophysical and computational tools to gain a deeper insight into the processes at the molecular level will be described.

1. Introduction

In the last decades, light has turned into one of the most relevant tools to activate and control chemical reactions. As an energy input, light offers the critical advantage of not generating any wastage, as most of the chemical or electrochemical inputs do. In addition, it can be extremely selective, in terms of both space and time. The recent development of powerful yet selective and cheap light sources has also contributed to the widespread of the use of light to provide the required energy for a chemical reaction. At the same time, light can also be used to monitor the progress of the reaction allowing, for instance, an excellent way of controlling the appearance of products. In parallel with these technical developments, the last forty years have also witnessed the birth and early development of new ideas on systems that could be activated and controlled by light. Undoubtedly, among the different uses of light on chemical systems, the most fascinating field is that of light-activated molecular machines. Machines have greatly contributed to the progress of humankind since the development of very simple tools to help in the everyday needs. The increase in complexity of tools and machines allowed the simplification of common tasks, but also to face new, more complex ones. The industrial revolution marked a milestone in the design and use of machines of increasing complexity and it is reasonable to say that we are on the verge of a new era that may be soon compared to the extreme breakthroughs that the industrial revolution produced. Although we are still starting to understand how to build and control efficient systems, molecular machines have already shown an immense potential in a wide range of different fields. This has been already recognized by the Nobel Prize in Chemistry awarded in 2016 to Jean-Pierre Sauvage, Sir James Fraser Stoddart and Bernard (Ben) L. Feringa by the Royal Swedish Academy of Sciences "*for the design and synthesis of molecular machines.*" Using a common and general definition, a molecular machine is an *assembly of a distinct number of molecular components that are designed to perform machinelike movements as a result of an appropriate external stimulation* (Balzani et al. 2000). In other words, one or several molecules or molecular parts working together to achieve some kind of motion.

When discussing molecular machines, two distinct factors are relevant: the stimulus that causes the machine motion (input) and the movement caused (output). The different inputs that can be used and the diverse outputs that can be obtained have been employed to classify molecular machines. In this sense, many different molecular assemblies have been designed and constructed and most of them imply a linear or rotational movement. Beyond the specific type of motion, the movement performed by any molecular machine is required to be controlled, that is, the motion should take place only when it is required (usually, when the stimulus is operating). Regarding the type of stimulus causing the movement, different systems triggered by a chemical reaction or an electrochemical input have been designed. However, due to the many practical advantages pointed out above, light is by far the most convenient stimulus to operate a molecular machine. The seminal works by Sauvage, Stoddart and Feringa have greatly influenced the field, but we will not review in detail their achievements. Instead, we will briefly describe their early findings to put in context the current state-of-the-art.

The first examples of efficient synthesis of compounds that now can fall into the category of molecular machines were due to the Sauvage. In the 1970s and 1980s, he designed the synthesis of two different types of compounds (i.e., rotaxanes and catenanes, see Figure X). These compounds were previously known but extremely difficult to prepare in adequate quantities for a detailed study. Using a copper atom as a coordination center, he and his team used a template synthesis to prepare rotaxanes and catenanes in good yields (Dietrich-Buchecker, Sauvage, and Kintzinger 1983). These allowed a detailed study of these new compounds and their unusual properties. Rotaxanes and catenanes are molecular assemblies in which the different parts are kept together by a singular type of interaction: the mechanical bond. This type of interaction allowed for the preparation of new molecules with increasingly complex structures (Sauvage and Weiss 1985) and boosted the field of topological chemistry (Nierengarten, Dietrich-Buchecker, and Sauvage 1994). As a consequence, the scientific community started to pay attention to the early efforts to design and build the previously anticipated molecular machines.

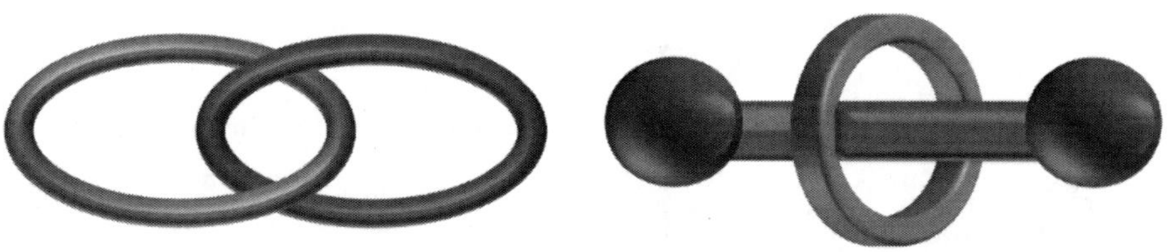

Figure 1. Cartoon representations of a rotaxane (left) and a catenane (right).

Both catenanes and rotaxanes showed an intriguing feature: the translational isomerism. Upon a certain stimulus, one part of the molecule could move reversibly with respect to the rest of the molecule. The first observation by Sauvage using a copper atom as chemical stimulus (Cesario et al. 1985) was soon followed by the exhaustive work of Stoddart on rotaxanes (Odell et al. 1988, Ballardini et al. 196). The stage was then set for the development of new systems and the preparation of new molecules while learning the different alternatives to efficiently control the motion of the molecular machines. Chemical, electrochemical and thermal inputs were tested successfully. However, it was soon revealed that the use of light to activate molecular machines allowed for a cleaner, easier and more efficient use. In this sense, the design of molecular machines with rotary movement based on overcrowded alkenes and fueled by light by Feringa resulted critical to understand some of the basic requirements for an efficient use and, most importantly, to use these systems in the plethora of applications that has been reported (Koumura et al. 1999, Delden, Wiel, Pollard, Vicario, Koumura, et al. 2005, Fletcher et al. 2005). We will discuss these compounds more in detail in section 3.2.

The pioneering work by Sauvage, Stoddart and Feringa opened the door for a new research field in the confluence of chemistry, physics and material science. In the last three decades, molecular machines have attracted an increasing amount of attention from many different perspectives. Interestingly, the synthesis using new methodologies, computational and experimental studies on the mechanisms of action and the clever design of applications have allowed for a truly multidisciplinary approach that it is one of the causes for the great evolution of the field. This is especially true when it comes to applications in biology or materials science, as no wastage products hamper the functioning of the switch as could take place with, for instance, a chemical input. The vast growth of the knowledge gathered and the specialized character of some of the applications, from optogenetics to liquid crystals, discourage a comprehensive review of the subject. Instead, in this contribution we will introduce some of the basic concepts of the design and use of molecular photoswitches.

This chapter is organized as follows. In the next section, we will review some of the basic features that are relevant in the design and operation of molecular photoswitches. This includes some examples of natural photoswitches that have been used as a source of inspiration for the design of new compounds and applications, some basic definitions and a classification of molecular switches. In the subsequent sections, we will describe several families of molecular photoswitches based on *E/Z* photoisomerizations. For each family of compounds, the basic features and some synthetic details will be presented. Then, we will focus on the photophysics, photochemistry and, especially, on the photoisomerization process for each type of compound.

2. Molecular Photoswitches: Basic Concepts

Among the many different types of molecular machines that have been proposed, molecular switches have been most extensively studied. The simple and manageable concept of a molecular switch, together with the easy access to some molecules acting as a switch, allowed for some relevant applications that further fueled the interest in these systems. As in the more general case of molecular machines, different stimuli have been used to operate molecular switches, but light has been found to be of more practical use.

In many cases, the synthetic molecular devices prepared have been designed after gaining inspiration form natural compounds. Guided by evolution, Nature offers excellent examples of very efficient devices. There is a big number of biomechanical natural machines that can be used as a model for the design and preparation of our synthetic analogues or alternatives. These molecular machines are nanometric equivalents of everyday machines and from these natural examples, we may learn the basics of their functioning to design new and improved versions to perform different functions or to fine tune the properties of these systems. Many examples of nanoscale devices are found in Nature. For instance, the photosynthetic machinery of green plants and other photosynthetic organisms converts more energy and synthesizes a greater amount of organic compounds than the complete chemical industry of humankind. In addition, this process is performed in an extremely clean and efficient way using inexpensive raw materials. Some examples of natural molecular devices appear in different essential biological processes including the transport of cations, the enzymatic synthesis of ATP (Metha et al. 1999), muscle contraction (Schliwa and Woehlke 2001) and photoreception (Baylor 1996).

In some cases, the biological processes in which these natural molecular machines are involved start with a quite simple photoisomerization step. Typical examples could be the process of vision (*E/Z* photoisomerization of the rhodopsin chromophore), the transport of cations (photoisomerization of bacteriorhodopsin), the luminescence in diverse organisms (emission quenching by photoisomerization of the chromophore in the green fluorescence protein) and the photoresponse in plants (photoisomerization of phytochrome). The photoisomerization of several natural chromophores has been recently reviewed (Gozem et al. 2017).

The complete biological process of vision is very complex and some details are not fully understood yet. However, the basic chemical step that initiates the biological cascade is extremely simple. Upon irradiation, the retinal chromophore undergoes a *E/Z* photoisomerization that is detected by the protein environment (rhodopsin) in which the chromophore is embedded (Kandori, Shichida, and Yoshizawa 2001). In the absence of visible light, the rhodopsin chromophore (a protonated Schiff base of the retinal) presents a *cis* conformation between the carbon atoms 11 and 12. When this chromophore absorbs a photon of visible light (λ between 400-600 nm), a very selective and fast (around 200 femtoseconds) photoisomerization of this double bond takes place through a π -> π^* singlet excited state (S_1) that decays to the all-*trans* ground state (S_0) photoproduct. This *cis-trans* photoisomerization of the chromophore is the primary chemical reaction in the process of vision and it is followed by a conformational change in the rhodopsin protein. This modification of the protein activates trasducin in order to initiate the biological process of vision (McBee et al. 2001). This *cis-trans* photoisomerization is highly efficient in rhodopsin, with an *in vivo* isomerization quantum yield (Φisom) of 0.67. This very high value allows for twilight vision. Due to this very high efficiency, the photoisomerization of the rhodopsin chromophore *in vivo* has been widely studied and used as a source of inspiration for the design of new prototypes of photoactivated molecular switches and motors capable of reproducing in solution the features of the chromophore inside the protein.

The chromophore of the green fluorescent protein (GFP) is another example of a photoisomerization process involved in a biological function.

Isolated from the *Aequorea victoria* jellyfish, GFP and other similar proteins feature fluorescence mediated by a photoexcitable green-light emitter chromophore (Heim, Prasher, and Tsien 1994). The chromophore responsible for this emission is (*Z*)-4-(4-hydroxybenzylidene)-1, 2-dimethyl-1*H*-imidazol-5(4*H*)-one, that it is formed by a phenolic ring and an imidazolidinone ring connected by a C=C double bond. It is embedded in a protein cavity and connected to it by a number of hydrogen bonds involving also some water molecules. The GFP chromophore is the responsible for the fluorescence of this protein in a picoseconds time scale that can be quenched by photoisomerization. Several studies have shown that the chromophore may also undergo a competitive non-radiative process upon light absorption. Thus, in this case the *E/Z* photoisomerization contributes to reduce the luminescence quantum yield (Voliani et al. 2008).

A large and rather heterogeneous family of biomolecules in which a relevant *E/Z* photoisomerization takes place is phytochromes. These photoreceptors are present in plants and in some types of bacteria and fungi (Quail et al. 1995). These proteins are formed by two main domains, namely the globular N-terminal domain (65 kDa), responsible for the binding to the chromophore, and the open C-terminal domain, mostly structural. The photoactive part is formed by a bilin chromophore (an open tetrapyrrolic system with the rings connected by methene moieties). This structure has an extended conjugation responsible for the absorption at long wavelengths. The chromophore may be interconverted between two different isomers by the effect of light. A dark adapted, red-absorbing isomer P_r and a metastable far-red-absorbing isomer P_{fr}. Upon light absorption (λ_{max} P_r = 660 nm), the system photoisomerizes along an exocyclic C=C double bond with a subsequent proton transfer step to form the photoisomer P_{fr} (λ_{max} = 730 nm). This basic photochemical step induces a reversible conformational change in the protein environment, which in turn initiates the phytochrome-signaling cascade. Either the P_r or P_{fr} isomers may be responsible for the biological function. In fact, the P_r/P_{fr} ratio controls the biological response, and this depends on the incident light and reaction kinetics of the photoisomerization and thermal back-reaction (Rockwell, Su, and Lagarias 2006).

Among the myriad of functions controlled by phytochromes, the seed germination process and the photomorphogenesis in plants, and the growth and pigmentation in bacteria could be mentioned. Chemically related to phytochromes, bilin chromophores can also be found in humans. They are degradation metabolic products of porphyrins (for instance, bilirubin). The photochemistry of bilirubin has been subject of attention mainly due to its influence in several diseases (high levels of bilirubin are responsible of neonatal jaundice). In all these chromophores, an exocyclic C=C double bond photoisomerizes from the dark stable *ZZ* isomer to the metastable *ZE* (McDonagh, Palma, and Lightner 1980).

As described in the introduction, a molecular machine can be defined as an assembly of several molecular parts that performs a controlled movement in response to an external stimulus. Thus, a molecular machine consumes the energy provided as input to execute a defined motion. More specifically, a molecular switch is a type of molecular machine that can be reversible interchanged between two different states. In this case, the input is used to provide the required energy to the molecule that eventually will lead to an output, a change in the geometry of the molecular switch. Using this definition, a molecular switch could be any molecule or molecular assembly whose geometry could be modified in a controlled fashion by means of an external input. When the energy source to be consumed is light (i.e., when the external input is light), the term molecular photoswitch applies. A molecular photoswitch should have, at least, two different geometries or states, and the conversion of one state to the other should be performed by light. The reverse conversion (from the final to the initial state) could be performed by light of a different wavelength, heat or a chemical stimulus (Figure 2).

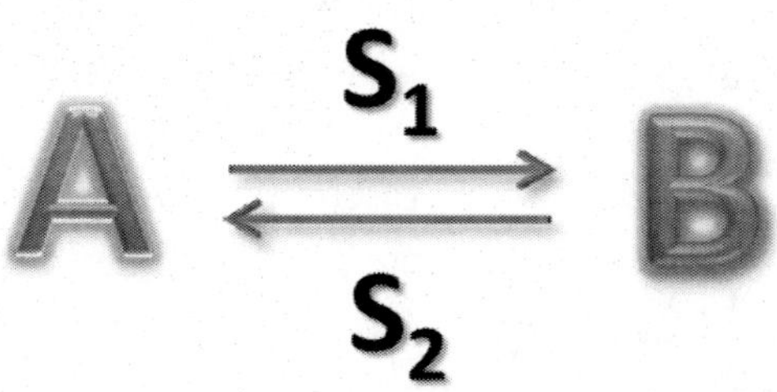

Figure 2. Schematic representation of a molecular switch. The initial stimulus (S_1) is light in a photoswitch. The stimulus used for the reversion (S_2) is usually either light or heat.

It is clear that many different types of chemical structures could fit into the general definition of a molecular photoswitch. In practice, two more requirements are implicitly used to term a molecule as a photoswitch: controlled motion and efficiency. With respect to the control of motion, for practical reasons only structures that could be interchanged between states in a controlled fashion are considered switches. This means that the change from state A to state B should take place exclusively when the stimulus is present, that is, when it is wanted to happen. In addition, this movement should be efficient, meaning that a reasonable fraction of the molecules should go from state A to state B when required. Many other features could be needed for specific applications: robustness, quickness, solubility… Thus, the specific chemical structure should be designed and prepared with all the required features in mind.

In the last decades, many different chemical structures have been designed to act as photoswitches. The geometrical motion caused by the change from state A to state B has been accomplished using very different moieties. Beyond some specific examples, most of the compounds fall into two main categories, namely photoswitches based in a cyclization and photoswitches base in *E/Z* isomerism. In this chapter, we will focus on the second category but, due to their relevance and interest, some brief introduction to the first type of photoswitches will be presented. Figure 3 illustrates some specific examples of photoswitches based on cyclizations. The application of this type of systems mostly rely in the difference in the optical or electronic properties of the isomers. The two states of the switch in this case are usually a cycle and an open form. These two isomers habitually feature different colors, which facilitate the conversion by light from one state to the other. This is connected with a change in a specific property: pKa, emission, magnetism… In contrast, the geometrical change induced by the change from state A to state B is minor.

Fulgides (Yokoyama) are a type of compounds consisting of a succinic anhydride moiety (or succinimide in the case of fulgimides) connected to an aromatic ring by a double C=C bond. Upon light absorption, the most stable, open (and usually colorless or yellow) isomer cyclizes to yield the closed isomer (yellow or green colored). As a recent example, fulgide derivatives have been used to photocontrol de biological activity of dopamine receptors (Lachmann et al. 2017).

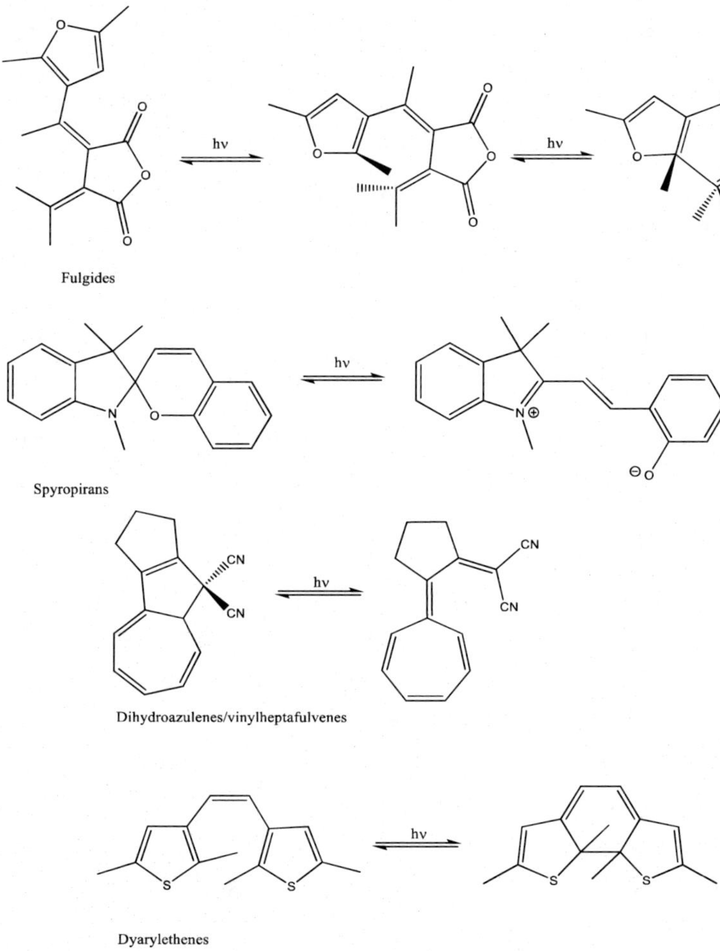

Figure 3. Typical examples of photoswitches based on cyclizations.

Spiropyrans (and the closely related spirooxazines) (Berkovic, Krongauz, and Weiss) are synthetically accessible and photocycle with high values of quantum efficiency. However, in general the metastable cycle reverts back to the initial open isomer with relative ease. A relevant recent application in which a spiropyran is used as a reversible photoacid has appeared (Halbritter et al. 2017).

The dihydroazulene/vinylheptafulvene couple (Vlasceanu et al. 2016) is an example of one-way photoswitch. Upon irradiation, the dihydroazulene cycle opens up to yield the metastable vinylheptafulvene that, eventually, will revert to the initial cycle, releasing the stored energy. This ability to store part of the light energy absorbed as chemical energy has found application in energy storage systems (Wang et al. 2017).

Among the photoswitches based on cyclization, diarylethenes are the compounds most extensively studied and used in applications. Due to this, a good number of reviews already exist (Tian and Wang 2007, Raymo and Tomasulo 2005, Janet, Erhan, and M. 2009, Yun et al. 2009). Some very relevant properties make these photoswitches adequate for applications in materials science. They show a very high fatigue resistance and both isomers (ring and open forms) are usually stable in the absence of light and feature relevant differences in their electronic and optical properties. Thus, these species have found different uses in the control of properties of films, liquid crystals, polymers and other materials.

Beyond the photoswitches based on cyclizations, another main category of photoswitches has been developed and found extensive application: photoswitches based on *E/Z* isomerizations. Light can be used to trigger the isomerization of certain type of double bonds. C=C, C=N and N=N are well-known chromophores that usually undergo isomerization upon light absorption. This property has been used to design and build molecular photoswitches that rely on the geometrical changes caused by light in chemical structures that include these double bonds. In the next section, we will review some of the compounds that have been reported. Some basic hints on the synthesis, together with the relevant properties and how to control them, will be discussed with a focus on the photoisomerization process. Unless explicitly stated, most of the following examples refer to direct excitation. Other ways to populate the excited states will not be explicitly covered here.

3. Molecular Photoswitches

3.1. Azobenzenes

Among the different structures that have been used to build molecular photoswitches, azobenzenes are the most used ones without a doubt. Many different properties contribute to their extensive use in diverse applications: azobenzenes feature strong absorptions in several regions of the UV-visible spectrum, photoisomerization is relatively efficient, the half-life of the photoisomer can be tuned from microsecond to days, the photoisomerization causes a big geometrical change… However, most of these properties can be achieved due to the extensive research that has been carried out on these compounds. The initial reasons that fueled the study of azobenzenes were more practical: azobenzenes have been known and used since more than a century and a half (Noble 1856), as they were key products in the beginnings of the chemical industry (Griffiths 1972). Azobenzenes were initially used as dyes in the textile industry and the requirements of easy and affordable synthesis together with the need for different properties caused an extensive research in the area. When their potential as photoswitches was revealed, the scientific community already knew a lot about azobenzenes, including synthetic routes to prepare many different compounds with diverse properties in good yields. In turn, this caused the rapid and widespread use of azobenzenes to photocontrol properties in a wide variety of systems. The popularity of azobenzenes as molecular photoswitches is reflected by the good number of reviews available, with the focus on the synthesis (Hamon et al. 2009), the photoisomerization and their use with biomolecules (Beharry and Woolley 2011) or as efficient photoswitches (Garcia-Iriepa et al. 2013). Also, several studies featuring the photocontrol by azobenzenes of different properties has appeared in the last years (Feringa and Browne 2011). In this contribution, we will focus on the photoisomerization properties, while we will briefly review the synthesis and properties. In brief, several straightforward synthetic routes to azobenzenes have been developed. For instance, the oxidative coupling or amines (Hirano et al.

1998), the reduction of nitro compounds (Kim et al. 2012) or the coupling of diazonium aromatic derivatives (Lee et al. 2007).

The isomerization mechanism in azobenzenes usually takes place when the *trans* isomer (in general the most stable isomer and the one obtained as the main product in most synthetic routes) absorbs light and forms the *cis* isomer as the main photoproduct. These compounds are quite photostable, so they can be irradiated for prolonged times with only minimal decomposition. The *cis* isomer is usually metastable and reverts back to the *trans* isomer in the dark. Due to the extensive study of azobenzenes and the availability of compounds with different substitution, most of the properties of azobenzenes can be tuned and may change very much from the unsubstituted azobenzene. For instance, the main absorption band in *trans*-azobenzenes can vary enormously, as it happens with the photostationary state, the half-life of the *cis* isomer or the solubility. All these changes on the chemical structures that also reflect on the properties may cause variations in the photoisomerization mechanism. This have caused considerable debate in the literature on the specific details of the photoisomerization of azobenzenes. Several mechanisms have been proposed to account for the experimental details of the photoisomerization of azobenzenes and different derivatives (see Figure 4).

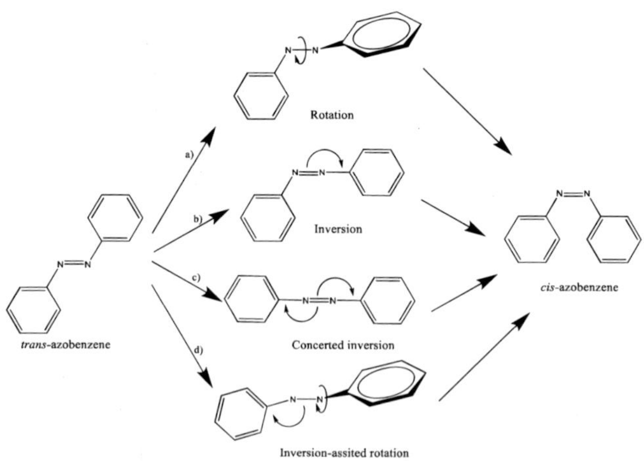

Figure 4. Different mechanisms for the *trans* to *cis* azobenzene photoisomerization. Adapted from (Bandara and Burdette 2012).

The rotation mechanism (Magee, Shand, and Eyring 1941) (Figure 4, a) implies the photochemical breaking of the N=N double bond to yield a freely rotating N-N bond. Thus, from the *trans* isomer with a dihedral angle of *ca.* 180º, upon light absorption the *cis* isomer with a dihedral angle of 0º may be formed. In the inversion mechanism (Curtin, Grubbs, and McCarty 1966) (Figure 4, b), the hybridization of only one of the nitrogen atoms in the N=N bond changes to *sp*. This causes the appearance of a transition structure in which that nitrogen atom has a linear disposition. Eventually, this structure would lead to the *cis* isomer. In the concerted inversion mechanism (Wei-Guang Diau 2004) (Figure 4, c), *both* nitrogen atoms are modified and the whole azobenzene adopts a linear disposition. Finally, in the inversion-assisted rotation mechanism (Tiberio et al. 2010) (Figure 4, d), a more complex movement is proposed, as a mixture of the pure inversion and the pure rotation movements. These four mechanisms have been discussed for the last decades and each of them claims to be based on experimental observations and/or computational data. However, a closer look to the conditions reveal that some key features may have not been taken into account or, at least, not considered to the same extent. Gas phase and solution measurements, different substitution patterns and diverse time scales and light sources for the irradiation are mixed along the literature, adding some noise to the underlying mechanism. Most probably, several (if not all) of the suggested mechanisms may be acting at the same time, and their contribution may be different depending on the experimental conditions. Thus, the main photoisomerization mechanism for a certain azobenzene under some experimental conditions may not be valid for a complete different situation. This should be taken into account when designing applications as far as the mechanism for the *trans*-to-*cis* isomerization may be relevant (for instance, in rigid systems when not all the postulated mechanisms could be possible).

The involved excited states and their impact in the photoisomerization quantum yields and the mechanisms has been also subject to debate. Both S_1 and S_2 have been reported to could take part in the *trans* to *cis* isomerization, while upper excited states are considered less relevant (Lednev et al. 1996). However, the mechanism of photoisomerization may be different when S_2

or S_1 are populated. In addition, T_1 and T_2 states have been computationally described to be energetically available, but no experimental confirmation of the population of these states upon direct irradiation has been found. In contrast, it was found that azobenzene could quench the triplet states of other species (Monti et al. 1981). In addition, other experimental conditions seem to have an impact on the mechanism: the wavelength of irradiation (as it could affect the excited state involved), the solvent (both polarity and viscosity influence the photoisomerization) and temperature (Bandara and Burdette 2012). Overall, many different factors contribute to a quite complex mechanistic picture. Even more, substitution of the azobenzene core may cause strong differences in the optical properties of the molecules with respect to the parent azobenzene. The involved excited states, the absorption bands and the thermal back reaction among other properties can be modified by substitution.

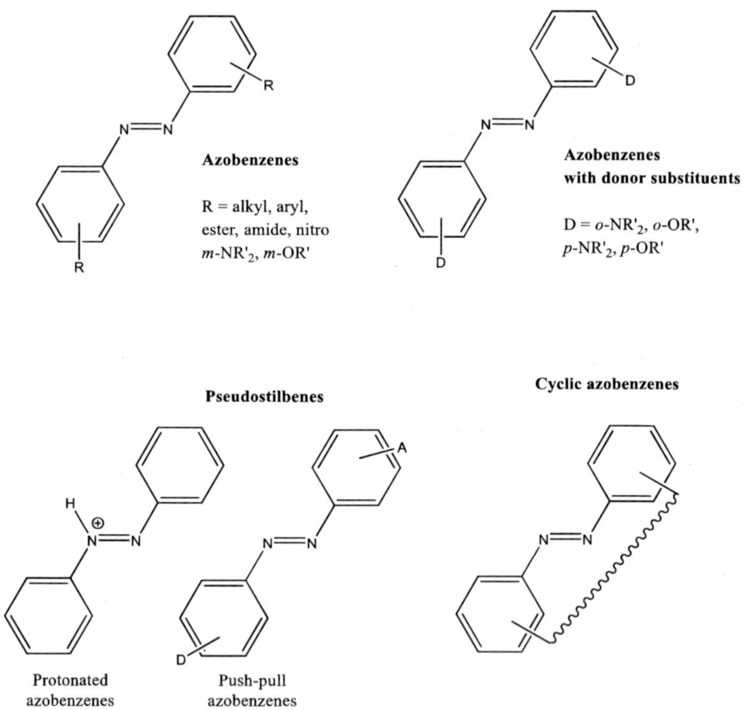

Figure 5. Types of azobenzenes based on their substitution.

Based on the chemical substitution of the basic core, several types of derivatives have been prepared and studied (Figure 5): azobenzenes, azobenzenes with donor substituents, pseudostilbenes and cyclic azobenzenes.

The first category of substituted azobenzenes include the parent molecule itself and those molecules with substituents that do not alter significantly the electronic properties of the basic core: substituents without a strong influence in the system (alkyl, aryl groups) or donor groups in meta position. For these compounds, the optical properties are very similar to azobenzene with a strong $\pi \rightarrow \pi^*$ absorption in the UV region and a somewhat weak $n \rightarrow \pi^*$ band, comparable quantum yields and relatively long half live for the *cis* isomer. Regarding the photoisomerization mechanism, substituent effects can alter the basic picture of the parent azobenzene, which implies an excitation to S_1, followed by relaxation in the excited state to reach a conical intersection and subsequent population of the ground state, each process in similar time scales. In this regard, restrictions in the relaxation of the molecule due to geometrical features or substitution may induce a variation in the mechanism from rotation to concerted inversion. Also, the thermal back-reaction of this type of azobenzene is very dependent on substitution (Dokic et al. 2009). For instance, electron-withdrawing substituents were found to decrease the energy barrier for thermal isomerization causing a reduction in the *cis* isomer lifetime. This has been extensively used to tune the stability of the photoisomers to match specific requirements of different applications.

The second main type of azobenzenes corresponds to those substituted by one or more donor substituents, typically amino or hydroxy groups in *meta* and/or *para* positions. For these compounds, the stronger $\pi \rightarrow \pi^*$ absorption is red-shifted and in some cases overlaps with the weaker $n \rightarrow \pi^*$ band. This shift has been used to trigger the photoisomerization by means of visible light. In addition, the donor substituents cause in general higher photoisomerization quantum yields as well as higher thermal stability (Ya et al. 2008). The presence of labile hydrogen atoms in the hydroxyl or amino groups may originate tautomerization processes and inter-molecular hydrogen bond interactions. In turn, this may originate aggregation-induced fluorescence and changes in the optical properties upon different pH values (Momotake and Arai 2003). The electronic effect of the donor substituents usually contributes to a relatively low in energy S_2 state which now becomes a bright state (in addition to S_1, in some cases). This also causes slower relaxation times and more complex photoisomerization mechanisms involving rotation, inversion or concerted inversion, depending on substitution. A special case of this type of compound are 2-hydroxyazobenzenes. The possibility to generate an intramolecular hydrogen bond, while usually does not impede photoisomerization, may cause significant differences in the optical and thermal properties such as low photoisomerization quantum yields and very fast thermal back-reactions (Bandara et al. 2011).

The third main class of azobenzene, namely pseudostilbenes, include protonated azobenzenes and push-pull systems. As in the donor-substituted azobenzenes, the $\pi \rightarrow \pi^*$ and $n \rightarrow \pi^*$ absorption bands overlap in the visible region. In the case of protonated azobenzenes, when a strong acid is used, the hydrogen atom is placed in one of the nitrogen atoms of the azo moiety, causing the $n\pi^*$ state to be higher in energy than the $\pi\pi^*$ excited state. This has not big effects in the photoisomerization, but causes a very fast thermal back-reaction. In turn, push pull systems include a strong electron-donating and a strong electron-withdrawing group in each of the *para* positions of the azobenzene. This special electron configuration originates the appearance of a new type of excited state related to charge transfer. These compounds usually absorb in the visible (they are strongly colored and used as dyes) and have very fast thermal back-reaction times (Marder et al. 1997). Photoisomerization in these compounds usually imply the low-lying $n\pi^*$ state and a rotation mechanism.

The last main class of azobenzenes are cyclic species. Most of the compounds than fall into this category are also substituted azobenzenes, as the ring chains are mainly alkyl substituents that do not have a big electronic influence into the conjugated π system. However, the restricted movement of these molecules imply that some of the possible mechanisms for the photoisomerization in equivalent open chain compounds are not available or have a very high energy barrier. For instance, rotation mechanism of photoisomerization is in some cases prevented due to geometrical restrictions (Rau 1984). However, the specific mechanism of photoisomerization is usually complex and can be affected by substitution beyond the ring confinement.

3.2. Overcrowded Alkenes

Apart from azobenzenes, photoswitches based on overcrowded alkenes have received much attention in the last two decades. This was highlighted by the Nobel Prize in Chemistry awarded in 2016 to Ben Feringa, one of the pioneers of the field. The main idea behind the use of this type of switches is the presence of a double C=C bond with substituents bulky enough to provide a big steric hindrance. As a result, the substituents around the double bond, initially planar, have to deviate from the plane of the molecule and arrange in a helical structure, as it happens with the family of compounds known as helicenes. If the two substituents in each of the two sides of the C=C double bond are different, these compounds become chiral. The synthesis and use of these compounds has been reviewed previously (Feringa et al. 2000, Feringa 2001, Feringa 2007). These compounds have found extensive use as part of chiroptical switches and motors (Cnossen, Browne, and Feringa 2014). The interested reader may appeal to some of the excellent reviews and books in the field for the synthesis and applications of overcrowded alkenes in biology, chemistry and materials science. In this contribution, we will briefly review some of the general features of the photoisomerization mechanism involved.

This type of molecules can appear in two main forms, namely *cis* and *trans*, due to the arrangements of the substituents around the C=C double bond. If the substituents are bulky enough to deviate from planarity, a chiral helical arrangement can take place, giving raise to two enantiomeric forms for each double bond configuration. The light induced isomerization between the *trans* and *cis* isomers is usually coupled with a thermal equilibration between diastereomers (see Figure 6).

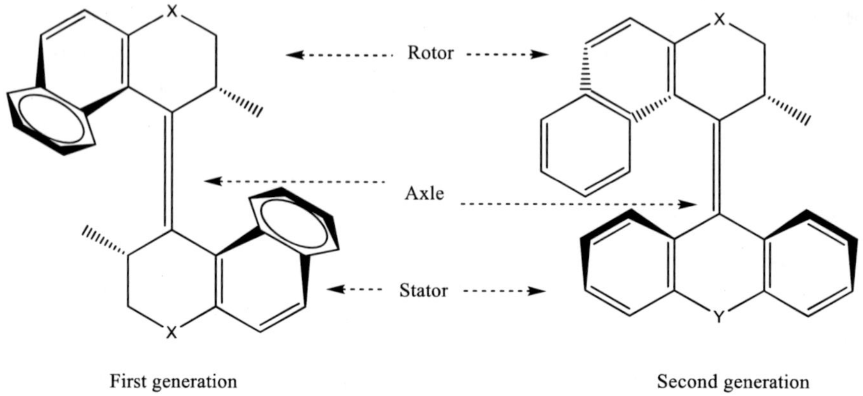

Figure 6. Chemical structures for the first- and second-generation of overcrowded alkene molecular motors.

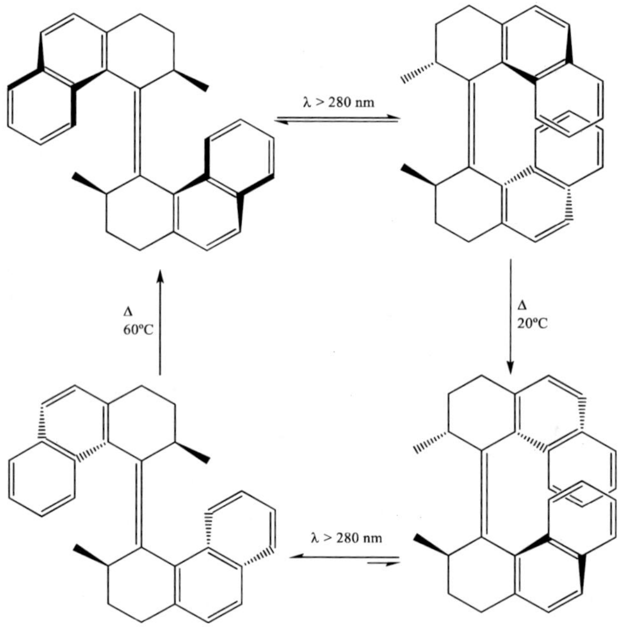

Figure 7. Typical four-step cycle for an overcrowded alkene molecular motor.

The design of overcrowded alkenes for use as light-driven molecular switches or motors has evolved through the years from very simple structures with intriguing, yet not very useful properties to elegant and practical systems. Chronologically, the first compounds of this class to be prepared (the so-called first generation) included the two same moieties in the "upper" (rotor) and "lower" (stator) halves, each one with one stereocenter (see Figure 6). In the second generation, the rotor and stator parts are different and only one stereocenter is present.

The basic photoisomerization mechanisms in both types is similar (see Figure 7). The central C=C double bond acts as the axle that keeps linked the two parts of the molecules. Upon light absorption, one part of the molecule (the rotor) is displaced with respect to the other part (the stator) though a stilbene-like *cis-trans* isomerization. The stereogenic elements present in the molecule (chiral atoms and helicity) originate a complex interplay responsible of the unidirectional movement of these molecules once light and/or thermal activation is applied. In most of the first general compounds, four different species are present in solution, interconnected by two photochemical steps (*cis-trans* photoisomerization) and two thermal steps (P,P and M,M isomers). While the photochemical steps are usually fast and can be activated by different wavelengths, the thermal steps induce a considerable delay in the unidirectional motion of the molecule. In some cases, the half-time for some steps is of the order of several days at 20ºC. This required heating of the mixtures for these compounds to be of practical interest. Thus, to activate these four-stage cycles, two different inputs are required. The light energy provides the energy to move the stable *cis* and *trans* isomers to their unstable stereoisomers, and the thermal energy is then used to surmount the barriers involved in the helix inversion. In turn, this thermal activation steps cause the unidirectional movement of the molecules. The thermal inversion of the helix can be controlled by substitution. The more hindered the alkene is (the more bulky the substituents) the slower the helix inversion will be. This fact has been used to study carefully every step of the cycle.

In the second generation of overcrowded alkenes, the good optical control of properties is maintained, while a speed up of the thermal steps is included through the design of asymmetric structures. Now, the two halves (rotor and stator, see Figure 6) are different and only one stereogenic carbon is present. This causes a lower energy barrier in the thermal isomerization steps (i.e., faster isomerization) and allows for different functionalization of the two parts of the molecule. This was used for the linking of these molecules to surfaces or nanoparticles (Delden, Wiel, Pollard, Vicario, and Feringa 2005). Once the light-induced unidirectional movement is controlled, and several other practical features may be tuned through design and synthesis, the major drawback of these molecules (the slow thermal steps) was addressed. As the photoisomerization steps usually are completed within picoseconds (Zijlstra et al. 1997), the overall speed of these molecular motors is limited by the thermal isomerization steps (which in the initial compounds take hours). In a series of different compounds prepared, this acceleration of the thermal steps was achieved through several improvements in the design. For instance, replacing the initial six-membered rings at the central C=C double bond by five-membered rings lead to a 10^5 acceleration of the helix inversion of one thermal step. In addition, changing the bridging atoms in either the rotor or the stator parts from the sulfur atom to a methylene unit further speeded up the isomerization. Again, this is related with the release of the steric hindrance in the C=C central double bond (Koumura et al. 2002). Steric effects of the substituents at the stereogenic atom may also contribute to tune the speed (Vicario et al. 2006). Increase the size of the substituent contributes to destabilize the unstable isomer, causing a decrease of the energy barrier and, thus, a faster isomerization. In addition, electronic effects may play a role. Push-pull systems were prepared to provoke the apparition of a resonance structure which could weaken the central C=C double bond, causing a faster isomerization (Kulago et al. 2010).

The fine-tuning of the thermal rotation steps allowed to prepare a full set of unidirectional photochemically activated motors with rotary motions speeds ranging from hours to microseconds. The specific compound to be used in a certain application should be selected based on this feature, but also in other more general properties: wavelength of irradiation, photochemical quantum yield, functionalization, effect of the environment. Fascinating applications of these molecules have been already designed (movement of macroscopic objects, molecular electronic devices, photoactive polymers, organization of liquid crystals…) and surely many more are yet to come.

3.3. Imines and Related Compounds

The *syn-anti* isomerization of imines (C=N double bonds) is known since the very early days of organic chemistry. However, their potential use as molecular switches or motors was achieved only in the 21st century. As in the case of alkenes, molecules containing C=N double bonds (imines, but also oximes, oxime esters and hydrazones) may appear in two (relatively) stable isomers, namely *E* and *Z*. Thus, this group of molecules could in principle be used as basic cores for the development of photoswitches and motors as it has been shown for overcrowded alkenes and azobenzenes. The C=N double bond could be isomerized with light as in alkenes, while the nitrogen atom could participate in a thermal inversion mechanism as azobenzenes. These two features could afford for the design of a light-activated unidirectional molecular motor (Lehn 2006).

With respect to the thermal *E/Z* isomerization of C=C double bonds, both the catalyzed and uncatalyzed versions have been explored. Clearly, the uncatalyzed version seems more practical, if under precise control, due to the ease of operation. If no catalysis is used, the energy barrier for the thermal isomerization has been shown to be very dependent on the nitrogen atom substituent, while substituents in the carbon atom show only little effect. For instance, when heteroatoms are directly linked to the iminic nitrogen (oximes, oxime esters), the resulting *E* and *Z* isomers are very stable due to the high energy barrier for interconversion. In contrast, aryl substituents on the nitrogen atom cause very fast thermal isomerization due to the low energy barriers. Between these two scenarios, alkyl substituents provide intermediate stability. Most of the experimental and computational studies support an inversion mechanism in the nitrogen atom, instead of a rotation mechanism.

The photochemical *E/Z* isomerization of imines and related compounds can be achieved by excitation to the singlet or triplet states and usually takes place *via* a rotational mechanism. The coupling of these two mechanisms (photochemical and thermal isomerism) could allow for a complete unidirectional cycle, the main requisite for a molecular motor (Greb and Lehn 2014).

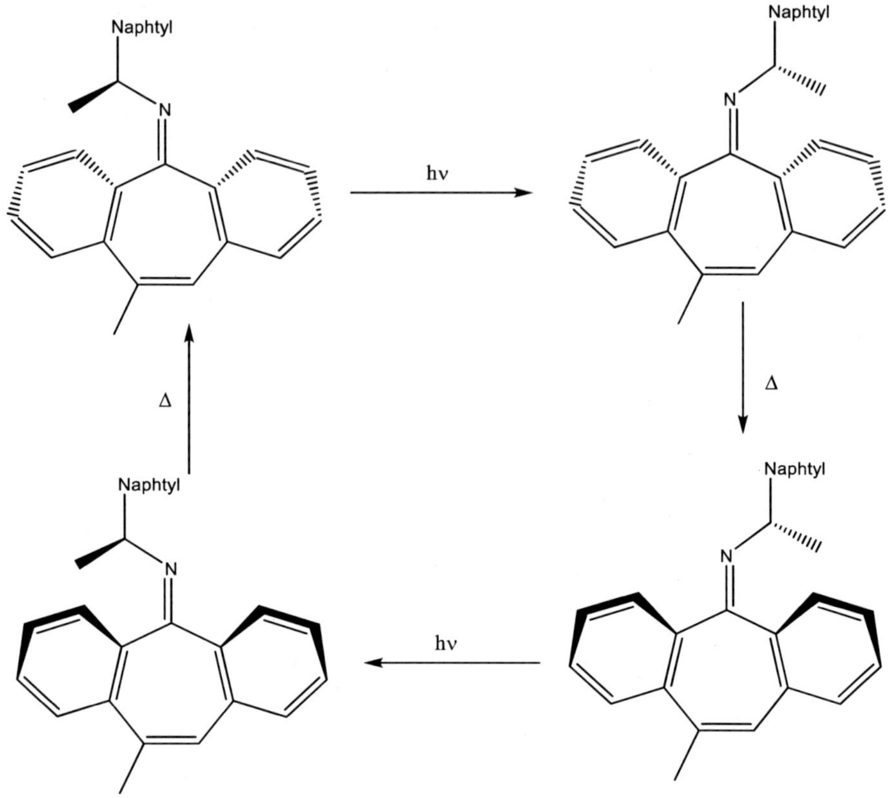

Figure 8. Four-step cycle on the photochemical and thermal isomerization of imine molecular motors. Adapted from (Greb and Lehn 2014).

Figure 8 shows the four stages cycle for one of the earliest examples of imine molecular motors. The initial (thermodynamically stable) diastereomeric mixture can be moved away from the equilibrium by using light. Eventually the initial composition can be recovered by thermal reaction (25-60ºC). While the photoisomerization takes place by a common rotation mechanism around the C=N double bond, the thermal isomerization could take place through a ring inversion (in a similar way to the helical isomerism in overcrowded alkenes) or through a nitrogen inversion (reminiscent to the azobenzene isomerism). Interestingly, these two mechanisms could be controlled my careful design of the molecules. When the seven-membered ring (the stator) is flexible enough, the conversion between the *E* and *Z* isomers takes place though a ring inversion. In contrast, with more rigid structures that increase the energy barrier for the ring inversion, the nitrogen inversion becomes favored.

In an effort to explore the smallest imine compound capable of behaving as a molecular motor, a series of camphor imines were prepared and studied by experimental and computational techniques (see Figure 9) (Greb, Eichhöfer, and Lehn 2015).

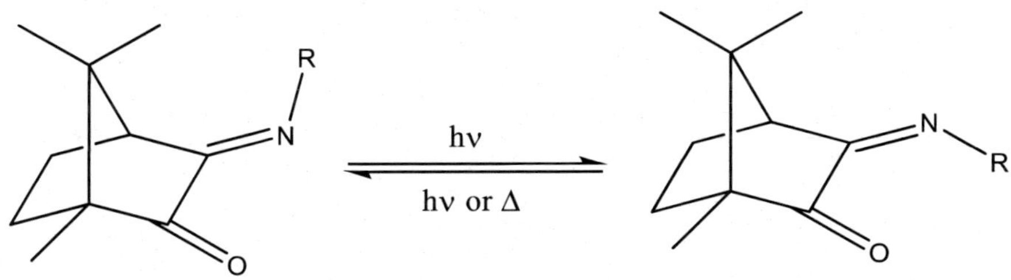

Figure 9. Camphor imine-based molecular motors.

For these very simple compounds, it was found that diastereoisomerism could drive the unidirectionality of the photochemical *E/Z* isomerism through a rotation mechanism, while the thermal part of the cycle can be accomplished by nitrogen atom inversion, thus originating a unidirectional motion. As a conclusion, any simple chiral imine could, in principle, behave as a molecular motor. In a subsequent study (Kulchat, Chaur, and Lehn 2017), authors also explored the possibility to merge this dynamic motion of imines with a dynamical constitution equilibrium between imines and their common components, aldehydes and amines, increasing the complexity but also the capabilities of the system.

3.4. Hemithioindigos

Hemithioindigos (HTIs) are asymmetric molecules composed by a thioindigo part connected to a stilbene part by a double bond, which allows the *Z*/*E* photoisomerization of the compound. These compounds may appear in two configurations: the *Z* configuration, which is thermodynamically more stable (Yamaguchi et al. 1992) and the *E* configuration, which is a metastable form (Reamonn and O'Sullivan 1977). Regarding the photochemical properties of the HITs, the *Z* form (λ_{max} = 440 nm) can be converted to the *E* form (λ_{max} = 470 nm) using light (>400 nm) which can be subsequently reverted to the *Z* form using light (500 nm) or thermally.

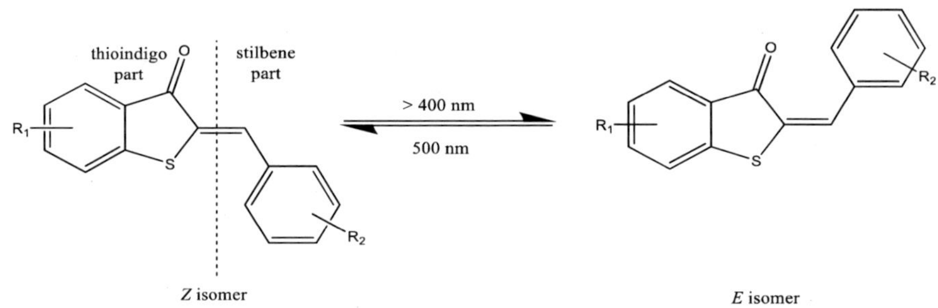

Figure 10. Hemithioindigo-based photoswitches.

HTIs feature a moderate photochromicity and in the absorption of the *E* isomer a bathochromic effect (20-30 nm) appears. This photochromicity allows to obtain the photoisomer (more than 95%) after irradiation. However, these compounds usually have a low quantum photoisomerization yield ($\phi_{Z/E}$ =14%-23%, $\phi_{E/Z}$ = 5-33%). Table 1 reports some specific features of HTIs in comparison with common azobenzenes.

The synthesis of HTIs can be done by different routes, although most usual ways are condensations reaction between aldehydes and benzothiophenones. The first one was described by Friedländer (Friedlander 1906) and consists of the formation of HTIs by condensation of a benzothiophenone and benzaldehydes, in which the ring of benzothiophenone was formed by an intramolecular condensation of 2 -[(carboxymethyl) thio] benzoic acid (see Figure 11, a).

Table 1. Comparison of the physical and photophysical properties of HTIs (Maerz et al. 2014) with azobenzenes (Bandara and Burdette 2012)

Compound	λ_{max} in nm (ε in $Lmol^{-1}cm^{-1}$)	Compositio n of PSS	Quantum yield ϕ	Excited state life time in ps	Activation barrier of thermal *E* to *Z* isomerization in $kcalmol^{-1}$
HTI *Z*	433 (12,600), 328 (17,000), 315 (19,800) in CH_2Cl_2	94% *E* (420 nm in CH_2Cl_2)	$\phi_{Z/E} = 0.23$ in CH_2Cl_2	$\tau_{Z/E} = 38$ in CH_2Cl_2	
HTI *E*	457 (5,600), 321 (13,500) in CH_2Cl_2	100% *Z* (505 nm in CH_2Cl_2)	$\phi_{E/Z} = 0.05$ in CH_2Cl_2	$\tau_{E/Z} = 23$ in CH_2Cl_2	$\Delta G^* = 31$ in toluene
Azobenzene *trans*	320 (21,300), 443 (500) in ethanol	90% *cis* (317 nm in benzene)	$\phi_{trans/cis}$ $\pi\pi^* = 0.11$ $n\pi^* = 0.25$ in n-hexane	$\tau_{trans} =$ 0.32; 2.1 in ethanol	
Azobenzene *cis*	281 (5,300), 432 (23,100) in ethanol		$\phi_{cis/trans}$ $\pi\pi^* = 0.44$ $n\pi^* = 0.56$ in n-hexane	$\tau_{cis} = 0.17$; 2.0 in ethanol	$\Delta G^* = 23$ in *n*-heptane

In 1992 Yamaguchi (Seki et al. 1992) synthesized a new family of HTIs by the route shown in Figure 11,f. Subsequently Mukherjee (Mukherjee and De 2002) developed a new route based on the deprotonation of the thiomethyl group in the *ortho* position of benzamides, followed by a ring closure (Figure 11,c). Later, this reaction was improved to a one step reaction using benzoates instead of benzamides by Cabiddu (Cabiddu et al. 2002). At the same time, in 2005 Knoieczny (Konieczny et al. 2005) developed a new synthesis via 4-acetyl-2-oxo-benz [1,3] oxathiole (Figure 11, d). Finally, in 2014, Wang (Zheng et al. 2014) proposed the formation of iodine mediated HTIs using *ortho*-methylthiophenyl vinyl ketones (Figure 11, e).

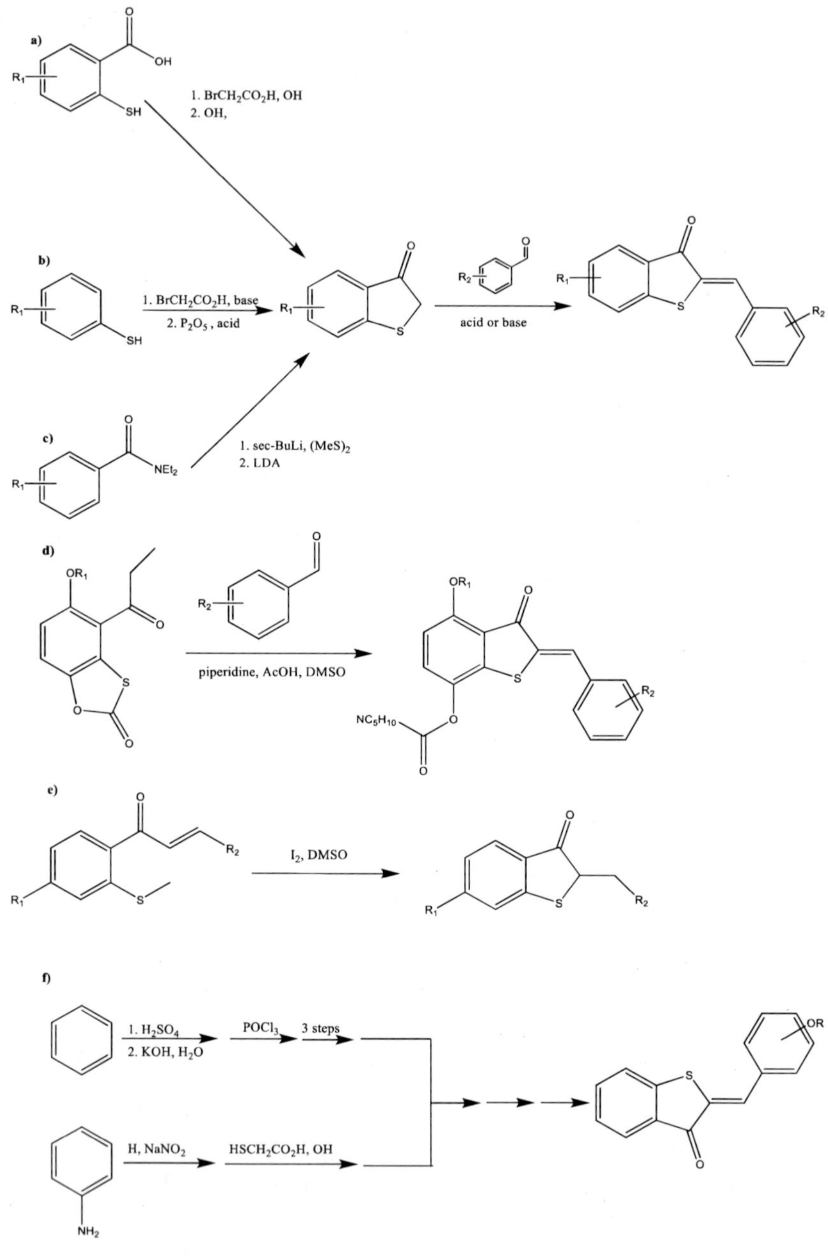

Figure 11. Syntheses of HTI. a) Friedlander's original synthesis; b) Intramolecular Friedel-Crafts acylation; c) Mukherjee synthesis; d) Domino reaction; e) Iodine mediated route; f) Yamaguchi synthesis.

The excited state of HTIs is usually a combination of two excited states (S_1 and S_2), which mix at a 90° rotation around the central C=C double bond. In the excited state potential energy surface, an energy barrier controls the speed of the *Z/E* photoisomerization. After surmounting this barrier, the molecule reaches a conical intersection, which connects the excited and the ground state (S_0). Upon light absorption and photoexcitation of the *Z* isomer in the ground state (S_0), population of the Franck-Condon region takes place, followed quickly (2-4 ps) by population of a local minimum. Torsion around the C=C central double bond leads to a biradical-like species formed at 90° rotation. For the *E* isomer the photoisomerization mechanism is similar with a virtually barrierless transition from the excited state to the ground state, which takes place before the double bond isomerization. This non-radiative "loss channel" is responsible for the low quantum yield of the *E/Z* photoisomerization as well as the very small fluorescence that is observed for the *E* isomer.

3.5. Benzylidene-Oxazolones

As already discussed above, Nature offers excellent examples of light-activated molecular switches. The GFP chromophore presented in the introduction is luminescent in a protein environment, but it becomes non-fluorescent when outside the protein barrel. This is due to a *Z*/*E* photoisomerization that opens up a non-radiative deactivation channel when the molecule is not subjected to the steric requirements of the protein. Based on the chemical structure of the GFP chromophore, molecules with the same benzylidene-oxazolone core were proposed as molecular photoswitches (see Figure 12) (Blanco-Lomas, Campos, and Sampedro 2012a). These compounds can be easily prepared in one-step from aldehydes with a different substitution pattern through a condensation mechanism. Irradiation of both *E* and *Z* isomers led efficiently to a photostationary state. The main isomer in the equilibrium can be controlled with a different stimulus, namely light and heat, or selection of the irradiation wavelength.

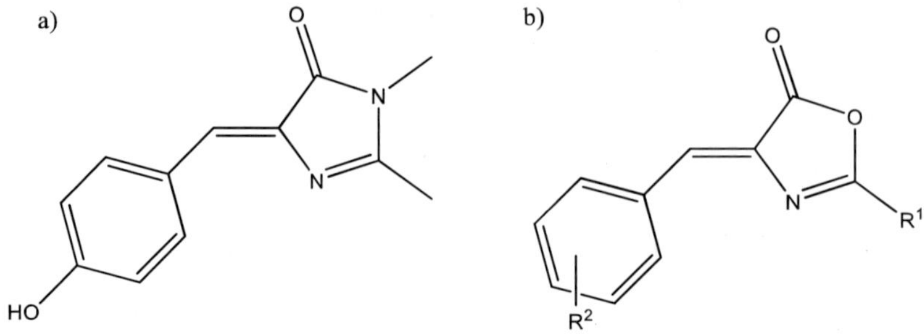

Figure 12. a) Green-fluorescent chromophore. b) Oxazolone-based molecular switches.

In a follow-up paper, the synthesis and photochemical properties of an extended family of oxazolones was reported (Blanco-Lomas et al. 2013). It was shown that the *E/Z* photoisomerization proceeds fast and efficiently and the photostationary state may be selected by different external factors such as the solvent, the type of stimulus (i.e., light or heat), the wavelength of irradiation but also by the substituents present in the oxazolone core. Thus, these variations allowed for a very precise control of the main isomer present in the equilibrium that could be relevant in order to fulfill the requirements of specific applications. These compounds showed a high thermal and photochemical stability. As a drawback, these species hydrolyze to some extent, which could prevent biological applications or the use of nucleophilic solvents.

A detailed computational study allowed to understand the photoisomerization mechanism (Funes-Ardoiz et al. 2013). Light absorption of relevant wavelengths lead to the population of both S_2 and S_1, as both excited states are very close in energy. Eventual relaxation to a minimum on the S_1 potential energy surface is in accordance with the observed fluorescence of these systems. However, the main deactivation pathway consist of a conical intersection located above the minimum in the excited state. The geometry of this conical intersection implies some changes with respect to the ground state. Both rings rotate and a kink is generated to obtain a typical space-saving structure typical of a hula-twist motion. Thus, while structurally related, the photoisomerization mechanisms of the GFP chromophore and benzylidene oxazolones differ clearly.

3.6. Arylidenehydantoins

Also related to the GFP chromophore, a different type of compounds based on arylidenehydantoins has been proposed as molecular photoswitches (Martínez-López et al. 2015). In this case, the imidazoline scaffold in the GFP is replaced by an imidazolinone moiety. While these compounds were known for quite a long time, their photochemical properties and their use as light-activated switches had been not evaluated. These species may be very easily prepared *via* condensation of hydantoin (or related reagents) with a variety of aldehydes. A different substitution pattern may be introduced at this stage to control some of the photophysical and photochemical properties of the switches. All the compounds prepared showed a very high thermal and photochemical stability, together with fatigue resistance, while the photoisomerization was found to be very dependent on the irradiation wavelength. For instance, in some cases no isomerization was found using a Pyrex filter to cut some of the wavelengths coming from a medium-pressure Hg lamp, while a quartz vessel allowed this isomerization to occur. This allows for an external tuning of the position of the switch, as quartz and Pyrex (i.e., different irradiation conditions) may control the switching process. In addition, the ratio of isomers in the photostationary state was found to be very dependent on the irradiation wavelength. Also, the photostationary state is greatly affected by the substituents in the aryl moiety (coming from the starting aldehyde) and thus it can be easily controlled.

A computational study allowed to understand the photoisomerization mechanism. After light absorption by the S_1 excited state of $\pi \rightarrow \pi^*$ character, the system evolves through relaxation in the excited state. The reaction coordinate is first dominated by stretching and subsequently by rotation around the C=C double bond. No minimum was found in the excited state in accordance with the lack of fluorescence and the fast photoisomerization process. After reaching a conical intersection point, the system can go back to the ground state with the formation of the photoproduct (i.e., the photoisomer) or the recovery of the starting isomer.

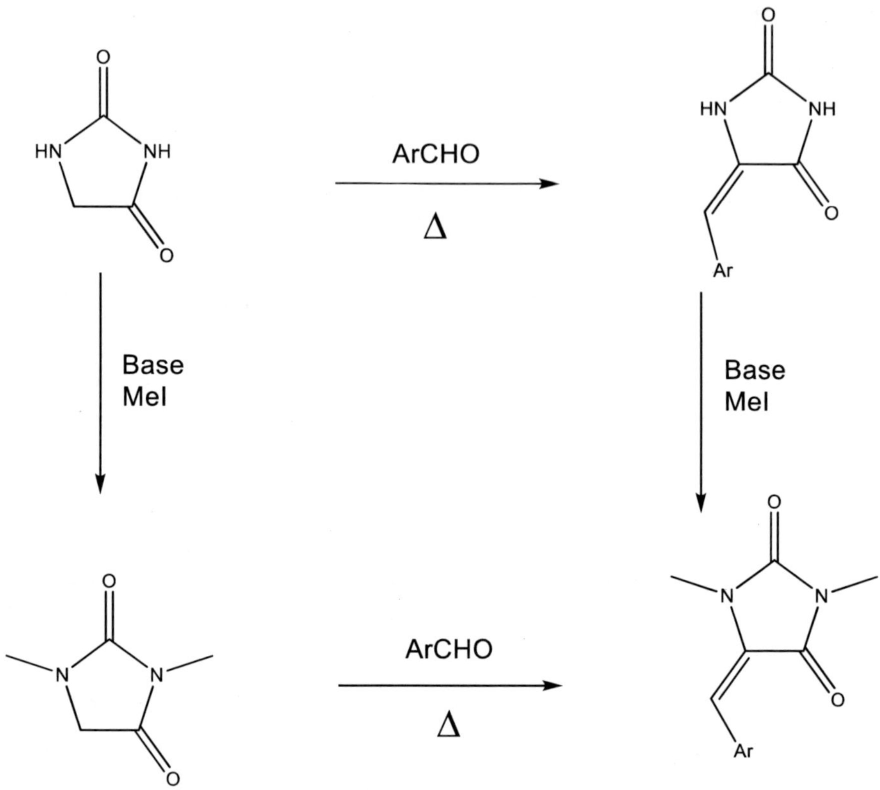

Figure 13. Synthesis of arylidenehydantoin-based molecular photoswitches. Adapted from (Martínez-López et al. 2015).

3.7. Arylpyrrolinones

Inspired by the relevant properties of the natural structure of the bilin and phytochrome chromophores, a set of molecules based on the basic core of these species has been suggested to act as molecular switches (García-Iriepa, Ernst, et al. 2016). While the whole molecule of the phytochrome chromophore includes four rings, the photoisomerization process in the native form takes place always in the same position, namely a C=C double bond within the dipyrrinone moiety (see Figure 14).

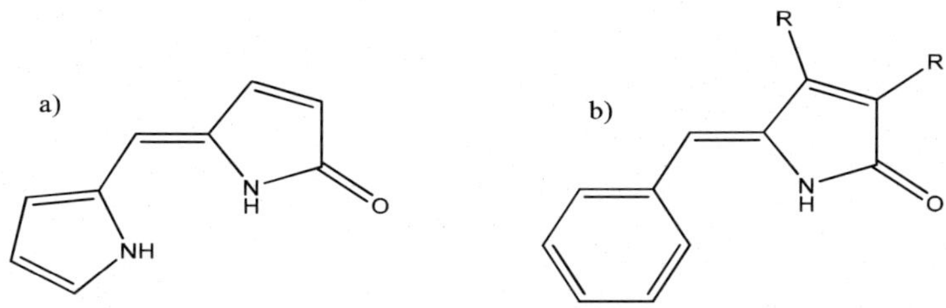

Figure 14. a) Dipyrrinone (from the phytochrome chromophore) and b) benzalpyrrolinone structures suggested to act as molecular photoswitches.

A series of compounds based on this basic core was prepared and their properties evaluated. Interestingly, both thermal (i.e., ground state) and photochemical properties could be changed through chemical modification. For instance, the configuration of the most stable isomer (*E* or *Z*) and the ability to form dimers was found to be dependent on the substituents present. In addition, the photoisomerization of the central C=C double bond was also found to be very dependent on the chemical structure. The most striking example of control of photoisomerization was found to be due to hydrogen bonding, which, when present, was able to completely impede photoisomerization.

A joint experimental and computational study allowed to explore in detail the mechanism of photoisomerization. Upon light absorption, the molecule is promoted to S_1, which was found to be the relevant excited state. Then, excited state relaxation takes place through two different time constants. The first, fast time scale, is related to the stretching of the molecule, which consists of the lengthening of the C=C central double bond. The second, slower time constant, implies the rotation around this double bond. No minimum was found in the excited state, in accordance with the fast isomerization process (also of the natural chromophore) and the lack of fluorescence. The relative decay times of the different compounds studied allowed to relate the speed of the photoisomerization processes with the shape of the potential energy surface involved and, in turn, with the chemical structure. Both electronic and steric effects caused by substitution may alter the photoisomerization dynamics. Methyl substitution in the central C=C double bond causes a relative weakening of the photoisomerizable bond, inducing a faster process. In addition, steric hindrance in the ground state caused by substituents may induce a pretwist of the C=C double bond, which also may accelerate the photoisomerization. The last part of the photoisomerization pathway is not dependent on substitution or the speed of the process. Eventually, the rotation along the C=C double bond in the excited state causes the apparition of a conical intersection point, which connects the excited state potential energy surface with the ground state. At this point, rotation along the C=C double bond is *ca.* 90º and decay to the ground state implies the formation of the photoproduct (either the *E* or the *Z* isomer depending on the initial isomer) and the partial recovery of the starting material.

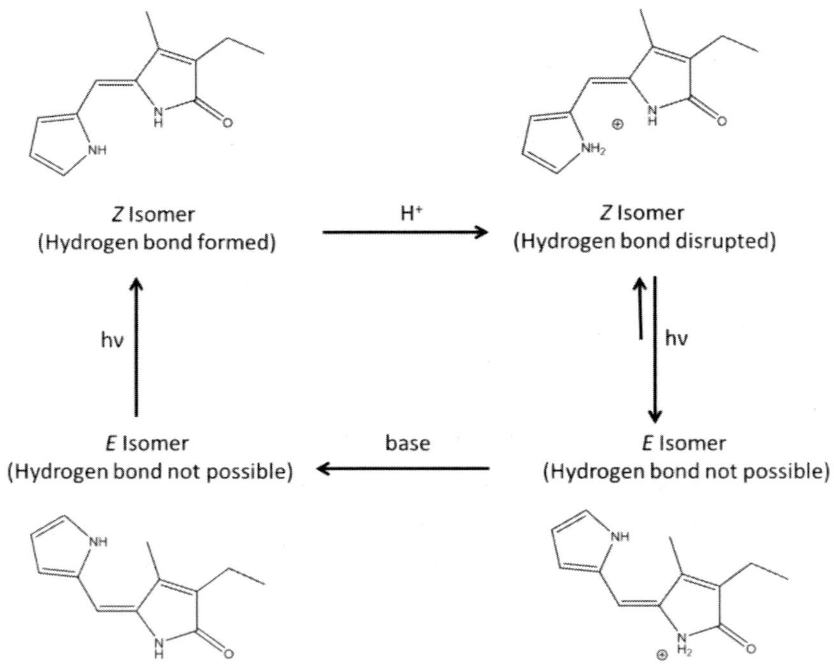

Figure 15. Four-stage cycle for an arylpyrrolinone molecular photoswitch. Adapted from (Martínez-López et al. 2015).

Interestingly, the interplay between the light-induced isomerization and the possibility to block this motion through a hydrogen bond allowed to prepare a four-step isomerization cycle. In this case, two different external inputs (light and medium acidity) were used to control the whole process. In the presence of acid, the hydrogen bond is disrupted, which in turn allows photoisomerization to take place. However, when a base is added to the system, the hydrogen bond is recovered and the photoisomerization in blocked. Thus, two different, orthogonal stimuli can be used to control the overall performance of the switch.

3.8. Retinal-Based Switches and Motors

As mentioned in the introduction, rhodopsins are a wide family of animal proteins found in very different species, which have in common their ability to provide a photoresponse. The specific properties and function depend on some subtle details that can vary from one type of organisms to other (Spudich 2006). However, these proteins share a similar core chromophore responsible for the interaction between protein and light. This chromophore is usually a protonated Schiff base of the aldehyde retinal. The specific chemical structure of the chromophore and the protein environment vary from one organism to other, which accounts for the different photophysics and photochemistry and, in turn, the function of the whole protein. Most of the biological functions mediated by these proteins are triggered by a quite simple chemical step: the light-activated isomerization of a C=C double bond (see Figure 16). The wide variety of functions in different organisms share a common photoisomerization step (with some differences from one function or type of organism to another) and in most cases, it is extremely efficient. As a typical example, the photoisomerization quantum yield for rhodopsin, the protein responsible for the primary process of vision, shows a quantum yield of 0.67 (Dartnall 1967). This very efficient photoisomerization has been used as inspiration in the design and preparation of a series of molecular photoswitches and motors. The synthesis, properties and applications of some of these systems have been already reviewed (Garcia-Iriepa et al. 2013).

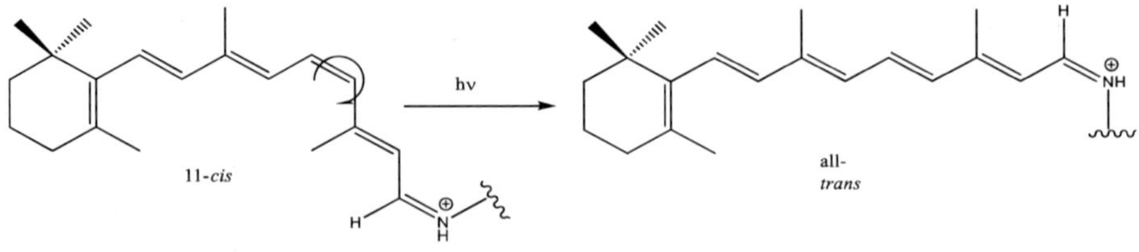

Figure 16. Photoisomerization of the protonated Schiff base of retinal.

Starting from the chemical structure of the rhodopsin chromophore, different approaches have been followed to prepare simple structures capable of performing a controlled motion upon irradiation (see Figure 17). The first attempts consisted on the truncation of the conjugated series of double bonds in the rhodopsin chromophore and the blocking of all these bonds but one (Sampedro et al. 2004). Doing so, the photoisomerization could be restricted to only one double bond. In a natural compound, this selectivity is achieved by embedding of the chromophore in the protein environment. These benzylidene-pyrrolines suffered for low quantum yields and a limited applicability. A new generation of compounds in which an additional cycle was included (indanylidenepyrrolines) in order to increase the photoisomerization quantum yield was prepared. Beyond the increased efficiency (0.3), these compounds offered fast isomerization and higher polarity (Sinicropi et al. 2008). As a drawback, the synthetic route for these compounds was long and tedious, and hardly adaptable for different substituents or applications. The addition of a new ring to the system (fluorenylidene-pyrrolines) allowed a further increase in the quantum yield (0.6) (Rivado-Casas et al. 2009). The preparation is simple and these compounds can be operated with visible light (Rivado-Casas, Campos, and Sampedro 2010). However, there is little control on the photostationary state composition through external stimuli (Rivado-Casas et al. 2011). Finally, a modification of the benzylidene-pyrroline switches (Blanco-Lomas, Campos, and Sampedro 2012b) could be prepared by a simple and versatile route, the photoisomerization could be achieved by low-energy light and the switch could be reset by heating. These systems could be applied in the photocontrol of the structure of a peptide (Blanco-Lomas et al. 2012).

Most of these systems share some common features in the photoisomerization mechanism, which, in turn, is based in the photoisomerization of the natural chromophore in rhodopsin. The protonated Schiff base (PBS) of the retinal in its different forms has been extensively studied by computational means (Lumento et al. 2007, Sampedro et al. 2004, Sinicropi et al. 2008), and this information has been applied to understand the behavior of the retinal-based switches but also in the design of new compounds. In most cases, these compounds are promoted to the S_1 state upon irradiation. When upper excited states are involved, the isomerization efficiency is usually lower. Then, the molecule starts relaxing in the excited state potential energy surface and exits the Franck-Condon region. This usually takes place through two different, sequential modes. The first one corresponds to a stretching mode in which the double and single bond order tend to invert. The C=C bond distances lengthen while the C-C distance shortens. In general, this stretching leads to a planar intermediate in the excited state in which the bond order of the extended chromophore (-C=C-C=C-) is inverted. From this planar intermediate in the excited state, a second molecular mode operates, corresponding with a torsional mode, which causes the isomerization of an initial C=C double bond. Through the stretching mode the C=C double bonds lengthen and turn into C-C which now are able to isomerize through the torsional mode. Eventually this leads to the *E/Z* photoisomerization. Recovery of the ground state takes place through a conical intersection in which the molecule (initially planar in the ground state) is now rotated *circa* 90° along the C=C double bond which isomerizes.

From this general picture, valid for the natural chromophore and some synthetic analogues, many different excited state properties have been modified or tuned to account for improved features. As stated above, the natural chromophore photoisomerizes very fast and effectively. The dynamics of the chromophore *in vivo* are controlled by the protein environment, which promotes and controls the isomerization by geometrical constrains. In contrast, the isolated chromophore (in either gas phase or solution) leads to a much slower *E/Z* isomerization. This is due to the lack of any constrains imposed by the protein. This allows the isomerization of different double bonds together with slow dynamics as the molecules are trapped in the planar excited state minimum.

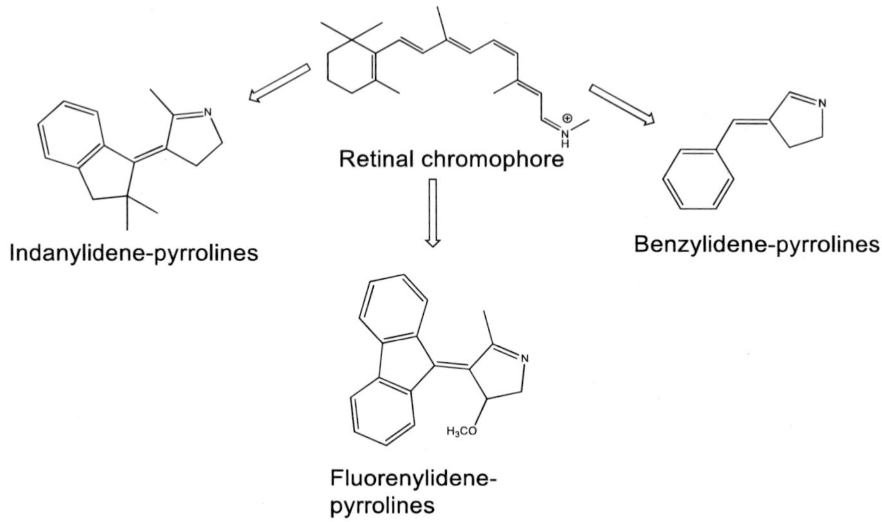

Figure 17. Retinal-based molecular photoswitches. Adapted from (Blanco-Lomas, Campos, and Sampedro 2012b).

Geometrical modifications of the basic chromophore (see Figure 17) allowed to approximate the properties of the molecules in solution to the values of the natural chromophore in a protein environment. For instance, the indanylidene-pyrroline switches feature a comparable isomerization quantum yield and excited state lifetime with respect to the rhodopsin (Lumento et al. 2007). The possibility to use visible light, compatible with biological and medical applications was achieved with the fluorenylidene-pyrroline switches. In this case, the extended π-system allows for a red shift of the absorption spectrum of *circa* 50 nm (Rivado-Casas et al. 2009). A further shift in the absorption accompanied with an increased photoisomerization quantum yield (0.64) was accomplished with the benzylidene-pyrroline switches (Blanco-Lomas, Campos, and Sampedro 2012b). Interestingly, these compounds are able to use visible light to alter the conformation of a peptide (Blanco-Lomas et al. 2012, García-Iriepa, Gueye, et al. 2016).

CONCLUSION

In this chapter, we have reviewed several systems that have been proposed to act as molecular switches based on the *E* / *Z* photoisomerism. Some of these compounds have found extensive use in many different disciplines from biology to materials science and will be studied more deeply in other chapters of this book. In contrast, some other species discussed here have been barely used in practical applications. This apparent inconsistency reflects, however, the possibilities that remain open in the field. A more systematic approach in the design and synthesis of a whole set of molecular switches will allow their use in many different applications. In addition, by selecting the adequate switch for a specific use, some of the drawbacks that are now present in some cases could be remove or, at least, mitigated. Even more, new and alternative uses of photoactivated molecular switches could be envisaged once compounds of very different properties (absorption, solubility, geometry change, photophysics…) will be available.

We have also briefly commented on the different mechanisms available for this type of compounds. An adequate knowledge of the underlying features of the functioning of the molecular photoswitches is required for the precise control of the switching process, especially once the compounds are incorporated into a complex system. The specific details of the mechanism of photoisomerization are markedly different from one type of switch to other. Even when the same basic scaffold is considered, clear differences between compounds can arise. Thus, a careful analysis of the factors governing the isomerization is probably necessary for a rational design and proper implementation of future successful applications. In this sense, the confluence of classic organic synthesis, theoretical calculations and photophysical studies has already proven to be a powerful combination.

The field of photochemical switches has already shown an impressive potential, turning from a chemical curiosity in the very early days to a potent tool to control many different features. Future development will surely offer new and exciting applications, but also a more precise control over many different properties.

Acknowledgments

Financial support from the Spanish MINECO (CTQ2017-87372-P) and ADER (2017-I-IDD-0003) is gratefully acknowledged.

References

Available from https://www.nobelprize.org/nobel_prizes/chemistry/laureates/2016/advanced.html.

Balzani, V., A. Credi, F. Raymo, and J. Stoddart. 2000. "Artificial Molecular Machines." *Angew. Chem. Int. Ed.* no. 39:3348-3391.

Ballardini, R., V. Balzani, A. Credi, M. T. Gandolfi, S. J. Langford, S. Menzer, L. Prodi, J. F. Stoddart, M. Venturi, and D. J. Williams. 196. "Simple Molecular Machines: Chemically Driven Unthreading and Rethreading of a [2]pseudorotaxane." *Angew. Chem. Int. Ed.* no. 35 (9):978-981.

Bandara, H. M., and S. C. Burdette. 2012. "Photoisomerization in different classes of azobenzene." *Chem. Soc. Rev.* no. 41 (5):1809-1825.

Bandara, H. M. D., S. Cawley, J. A. Gascon, and S. C. Burdette. 2011. *Eur. J. Org. Chem.*:2916-2919.

Baylor, D. 1996. *Proc. Natl. Acad. Sci. USA* no. 93:560-565.

Beharry, Andrew A., and G. Andrew Woolley. 2011. "Azobenzene photoswitches for biomolecules." *Chem. Soc. Rev.* no. 40 (8):4422-4437.

Berkovic, G., V. Krongauz, and V. Weiss. *Chem. Rev.* no. 100:1741-1754.

Blanco-Lomas, M., P. J. Campos, and D. Sampedro. 2012a. "Benzylidene-Oxazolones as Molecular Photoswitches." *Org. Lett.* no. 14 (17):4334-4337.

Blanco-Lomas, M., P. J. Campos, and D. Sampedro. 2012b. "Synthesis and Photoisomerization of Rhodopsin-Based Molecular Switches." *Eur. J. Org. Chem.* no. 2012 (32):6328-6334.

Blanco-Lomas, M., I. Funes-Ardoiz, P. J. Campos, and D. Sampedro. 2013. "Oxazolone-Based Photoswitches: Synthesis and Properties." *Eur. J. Org. Chem.* no. 2013 (29):6611-6618.

Blanco-Lomas, M., S. Samanta, P. J. Campos, G. A. Woolley, and D. Sampedro. 2012. *J. Am. Chem. Soc.* no. 134:6960-6963.

Cabiddu, M. G., S. Cabiddu, E. Cadoni, S. De Montis, C. Fattuoni, S. Melis, and M. Usai. 2002. *Synthesis*:875-878.

Cesario, M., C. Dietrich-Buchecker, J. Guilhem, C. Pascard, and J. P. Sauvage. 1985. "Molecular Structure of a Catenand and Its Copper(I) Catenate: Complete Rearrangement of the Interlocked Macrocyclic Ligands by Complexation." *J. Chem. Soc. Chem. Commun.*:244-247.

Cnossen, A., W. R. Browne, and B. L. Feringa. 2014. "Unidirectional Light-Driven Molecular Motors Based on Overcrowded Alkenes." In *Molecular Machines and Motors: Recent Advances and Perspectives*, edited by Alberto Credi, Serena Silvi and Margherita Venturi, 139-162. Cham: Springer International Publishing.

Curtin, D. Y., E. J. Grubbs, and C. G. McCarty. 1966. "Uncatalyzed syn-anti Isomerization of Imines, Oxime Ethers, and Haloimines1." *J. Am. Chem. Soc.* no. 88 (12):2775-2786.

Dartnall, H. J. A. 1967. *Vision Res.* no. 8:339-358.

Delden, R. A. van, M. K. J. ter Wiel, M. M. Pollard, J. Vicario, and B. L. Feringa. 2005. *Nature* no. 437:1337-1340.

Delden, R. A. van, M. K. J. ter Wiel, M. M. Pollard, J. Vicario, N. Koumura, and B. L. Feringa. 2005. "Unidirectional Molecular Motor on a Gold Surface." *Nature* no. 437 (7063):1337-1340.

Dietrich-Buchecker, C. O., J. P. Sauvage, and J. P. Kintzinger. 1983. "Une nouvelle famille de molécules: Les métallo-caténanes." *Tetrahedron Lett.* no. 24 (46):5095-5098.

Dokic, J., M. Gothe, J. Wirth, M. V. Peters, J. Schwarz, S. Hecht, and P. Saalfrank. 2009. *J. Phys. Chem. A* no. 113:6763-9773.

Feringa, B. L. 2001. *Acc. Chem. Res.* no. 34:504-513.

Feringa, B. L., R. A. van Delden, N. Koumura, and E. M. Geertsema. 2000. *Chem. Rev.* no. 100:1789-1816.

Feringa, Ben L. 2007. "The Art of Building Small: From Molecular Switches to Molecular Motors." *J. Org. Chem.* no. 72 (18):6635-6652.

Feringa, Ben L., and Wesley R. Browne. 2011. *Molecular Switches*: Wiley-VCH.

Fletcher, S. P., F. Dumur, M. M. Pollard, and B. L. Feringa. 2005. "A Reversible, Unidirectional Molecular Rotary Motor Driven by Chemical Energy." *Science* no. 310 (5745):80-82.

Friedlander, P. 1906. *Chem. Ber.* no. 39:1060-1066.

Funes-Ardoiz, Ignacio, Marina Blanco-Lomas, Pedro J. Campos, and Diego Sampedro. 2013. "Benzylidene–oxazolones as photoswitches: photochemistry and theoretical calculations." *Tetrahedron* no. 69 (46):9766-9771.

García-Iriepa, Cristina, Hanna A. Ernst, Yu Liang, Andreas-Neil Unterreiner, Luis Manuel Frutos, and Diego Sampedro. 2016. "Study of Model Systems for Bilirubin and Bilin Chromophores: Determination and Modification of Thermal and Photochemical Properties." *J. Org. Chem.* no. 81 (15):6292-6302.

García-Iriepa, Cristina, Moussa Gueye, Jérémie Léonard, David Martínez-López, Pedro J. Campos, Luis Manuel Frutos, Diego Sampedro, and Marco Marazzi. 2016. "A biomimetic molecular switch at work: coupling photoisomerization dynamics to peptide structural rearrangement." *PCCP* no. 18 (9):6742-6753.

Garcia-Iriepa, Cristina, Marco Marazzi, Luis Manuel Frutos, and Diego Sampedro. 2013. "E/Z Photochemical switches: syntheses, properties and applications." *RSC Advances* no. 3 (18):6241-6266.

Gozem, Samer, Hoi Ling Luk, Igor Schapiro, and Massimo Olivucci. 2017. "Theory and Simulation of the Ultrafast Double-Bond Isomerization of Biological Chromophores." *Chem. Rev.* no. 117 (22):13502-13565.

Greb, L., and J. M. Lehn. 2014. *J. Am. Chem. Soc.* no. 136:13114-13117.

Greb, Lutz, Andreas Eichhöfer, and Jean-Marie Lehn. 2015. "Synthetic Molecular Motors: Thermal N Inversion and Directional Photoinduced C=N Bond Rotation of Camphorquinone Imines." *Angew. Chem. Int. Ed.* no. 54 (48):14345-14348.

Griffiths, J. 1972. *Chem. Soc. Rev.* no. 1:481-493.

Halbritter, Thomas, Christoph Kaiser, Josef Wachtveitl, and Alexander Heckel. 2017. "Pyridine–Spiropyran Derivative as a Persistent, Reversible Photoacid in Water." *J. Org. Chem.* no. 82 (15):8040-8047.

Hamon, Florian, Florence Djedaini-Pilard, Francis Barbot, and Christophe Len. 2009. "Azobenzenes—synthesis and carbohydrate applications." *Tetrahedron* no. 65 (49):10105-10123.

Heim, R., D. C. Prasher, and R. Y. Tsien. 1994. *Proc. Natl. Acad. Sci. U.S.A.* no. 91:12501-12504.

Hirano, Masao, Sigetaka Yakabe, Hideki Chikamori, James H. Clark, and Takashi Morimoto. 1998. "Oxidation by Chemical Manganese Dioxide. Part 3.1 Oxidation of Benzylic and Allylic Alcohols, Hydroxyarenes and Aminoarenes." *J. Chem. Res., S.* (12):770-771.

Janet, Cusido, Deniz Erhan, and Raymo Françisco M. 2009. "Fluorescent Switches Based on Photochromic Compounds." *Eur. J. Org. Chem.* no. 2009 (13):2031-2045.

Kandori, H., Y. Shichida, and T. Yoshizawa. 2001. *Biochem. (Moscow)* no. 66:1197-1209.

Kim, J. H., J. H. Park, Y. K. Chung, and K. H. Park. 2012. *Adv. Synth. Catal.* no. 354:2412-2418.

Konieczny, M. T., W. Konieczny, S. Wolniewicz, K. Wierzba, Y. Suda, and P. Sowinski. 2005. *Tetrahedron* no. 61:8648-8655.

Koumura, N., E. M. Geertsema, M. B. van Gelder, A. Meetsma, and B. L. Feringa. 2002. *J. Am. Chem. Soc.* no. 124 (5037-5051)

Koumura, N., R. W. J. Zijlstra, R. A. van Delden, N. Harada, and B. L. Feringa. 1999. "Light-Driven Monodirectional Molecular Rotor." *Nature* no. 401 (6749):152-155.

Kulago, A. A., M. Klok, E. M. Mes, A. M. Brouwer, and B. L. Feringa. 2010. *J. Org. Chem.* no. 75:666-679.

Kulchat, Sirinan, Manuel N. Chaur, and Jean-Marie Lehn. 2017. "Kinetic Selectivity and Thermodynamic Features of Competitive Imine Formation in Dynamic Covalent Chemistry." *Chem. – Eur. J.* no. 23 (46):11108-11118.

Lachmann, D., C. Studte, B. Männel, H. Hübner, P. Gmeiner, and B. König. 2017. "Photochromic Dopamine Receptor Ligands Based on

Dithienylethenes and Fulgides." *Chem. – Eur. J.* no. 23 (54):13423-13434.

Lednev, Igor K., Tian-Qing Ye, Ronald E. Hester, and John N. Moore. 1996. "Femtosecond Time-Resolved UV−Visible Absorption Spectroscopy of trans-Azobenzene in Solution." *J. Phys. Chem.* no. 100 (32):13338-13341.

Lee, M. H., B. K. Cho, J. Yoon, J. S. Kim, *Org. Lett.,* and 9 2007, 4515. 2007. *Org. Lett.* no. 9:4515-4518.

Lehn, J. M. 2006. *Chem. – Eur. J.* no. 12:5910-5915.

Lumento, F., V. Zanirato, S. Fusi, E. Busi, L. Latterini, F. Elisei, A. Sinicropi, T. Andruniów, N. Ferré, R. Basosi, and M. Olivucci. 2007. *Angew. Chem., Int. Ed.* no. 46:414-420.

Maerz, B., S. Wiedbrauk, S. Oesterling, E. Samoylova, A. Nenov, P. Mayer, R. de Vivie-Riedle, W. Zinth, and H. Dube. 2014. *Chem. Eur. J.* no. 20:13984-13992.

Magee, John L., William Shand, and Henry Eyring. 1941. "Non-adiabatic Reactions. Rotation about the Double Bond." *J. Am. Chem. Soc.* no. 63 (3):677-688.

Marder, S. R., B. Kippelen, A. K. Y. Jen, and N. Peyghammbarian. 1997. *Nature* no. 388:845-851.

Martínez-López, David, Meng-Long Yu, Cristina García-Iriepa, Pedro J. Campos, Luis Manuel Frutos, James A. Golen, Sivappa Rasapalli, and Diego Sampedro. 2015. "Hydantoin-Based Molecular Photoswitches." *J. Org. Chem.* no. 80 (8):3929-3939.

McBee, J. K., K. Palczewski, W. Baehr, and D. R. Pepperberg. 2001. *Prog. Retin. Eye Res.* no. 20:469-529.

McDonagh, A. F., L. A. Palma, and D. A. Lightner. 1980. *Science* no. 208:145-151.

Metha, A. D., M. Rief, J. A. Spudich, D. A. Smith, and R. M. Simmons. 1999. *Science* no. 283:1689-1695.

Momotake, A., and T. Arai. 2003. *Tetrahedron Lett.* no. 44:7277-7280.

Monti, Sandra, Enrico Gardini, Pietro Bortolus, and Edmond Amouyal. 1981. "The triplet state of azobenzene." *Chem. Phys. Lett.* no. 77 (1):115-119.

Mukherjee, C., and A. De. 2002. *Synlett*: 325-327.

Nierengarten, J. F., C. O. Dietrich-Buchecker, and J. P. Sauvage. 1994. "Synthesis of a Doubly Interlocked [2]-Catenane." *J. Am. Chem. Soc.* no. 116:375-376.

Noble, A. 1856. *Justus Liebigs Ann. Chem.* no. 98:253-256.

Odell, B., M. V. Reddington, A. M. Z. Slawin, N. Spencer, J. F. Stoddart, and D. J. Williams. 1988. "Cyclobis(paraquat-P-Phenylene). A Tetracationic Multipurpose Receptor." *Angew. Chem. Int. Ed.* no. 27 (11):1547-1550.

Quail, P. H., M. T. Boylan, B. M. Parks, T. W. Short, Y. Xu, and D. Wagner. 1995. *Science* no. 268:675-680.

Rau, H. 1984. *J. Photochem.* no. 26:221-225.

Raymo, Francisco M., and Massimiliano Tomasulo. 2005. "Electron and energy transfer modulation with photochromic switches." *Chem. Soc. Rev.* no. 34 (4):327-336.

Reamonn, L. S. S., and W. I. O'Sullivan. 1977. *J. Chem. Soc. Perkin Trans. 1*:1009-1012.

Rivado-Casas, L., M. Blanco-Lomas, P. J. Campos, and D. Sampedro. 2011. *Tetrahedron* no. 67:7570-7574.

Rivado-Casas, L., P. J. Campos, and D. Sampedro. 2010. *Organometallics* no. 29:3117-3124.

Rivado-Casas, L., D. Sampedro, P. J. Campos, S. Fusi, V. Zanirato, and M. Olivucci. 2009. *J. Org. Chem.* no. 74:4666-4674.

Rockwell, N. C., Y. S. Su, and J. C. Lagarias. 2006. *Annu. Rev. Plant Biol.* no. 57:837-868.

Sampedro, D., A. Migani, A. Pepi, E. Busi, R. Basosi, L. Latterini, F. Elisei, S. Fusi, F. Ponticelli, V. Zanirato, and M. Olivucci. 2004. *J. Am. Chem. Soc.* no. 126:9349-9359.

Sauvage, J. P., and J. Weiss. 1985. "Synthesis of Dicopper(I) [3]Catenates: Multiring Interlocked Coordinating Systems." *J. Am. Chem. Soc.* no. 107:6108-6110.

Schliwa, M., and G. Woehlke. 2001. *Nature 2001, 411, 424.* no. 411:424-426.

Seki, T., T. Tamaki, T. Yamaguchi, and K. Ichimura. 1992. *Bull. Chem. Soc. Jpn.* no. 65:957-963.

Sinicropi, A., E. Martin, M. Ryasantsev, J. Helbing, J. Briand, D. Sharma, J. Léonard, S. Haacke, A. Canizzo, M. Chergui, V. Zanirato, S. Fusi, F. Santoro, R. Basosi, N. Ferré, M. Olivucci, and Proc. Natl. Acad. Sci. USA. 2008. *Proc. Natl. Acad. Sci. USA* no. 105:17642-17947.

Spudich, J. L. 2006. *Trends Microbiol.* no. 14:480-487.

Tian, He, and Sheng Wang. 2007. "Photochromic bisthienylethene as multi-function switches." *Chem. Comm.* (8):781-792.

Tiberio, G., L. Muccioli, R. Berardi, and C. Zannoni. 2010. "How Does the Trans–Cis Photoisomerization of Azobenzene Take Place in Organic Solvents?" *Chem Phys Chem* no. 11 (5):1018-1028.

Vicario, J., M. Walko, A. Meetsma, and B. L. Feringa. 2006. *J. Am. Chem. Soc.* no. 128 (5127-5135)

Vlasceanu, A., S. L. Broman, A. S. Hansen, A. B. Skov, M. Cacciarini, A. Kadziola, H. G. Kjaergaard, K. V. Mikkelsen, and M. B. Nielsen. 2016. "Solar Thermal Energy Storage in a Photochromic Macrocycle." *Chem., A Eur. J.* no. 22 (31):10796-10800.

Voliani, V., R. Bizzarri, R. Nifosì, S. Abbruzzetti, E. Grandi, C. Viappiani, F. Beltram, and 112 *J. Phys. Chem. B* 2008, 10714;. 2008. *J. Phys. Chem. B* no. 112:1071410-722.

Wang, Zhihang, Jonas Udmark, Karl Börjesson, Rita Rodrigues, Anna Roffey, Maria Abrahamsson, Mogens Brøndsted Nielsen, and Kasper Moth-Poulsen. 2017. "Evaluating Dihydroazulene/Vinylheptafulvene Photoswitches for Solar Energy Storage Applications." *Chemsuschem* no. 10 (15):3049-3055.

Wei-Guang Diau, Eric. 2004. "A New Trans-to-Cis Photoisomerization Mechanism of Azobenzene on the S1(n,π*) Surface." *The Journal of Physical Chemistry A* no. 108 (6):950-956.

Ya, Q., X. Z. Dong, W. Q. Chen, X. M. Duan, Dyes Pigm., and 79 2008, 159–165. 2008. *Dyes Pigm.* no. 79:159-165.

Yamaguchi, T., T. Seki, T. Tamaki, and K. Ichimura. 1992. *Chem. Soc. Jpn.* no. 65:949-656.

Yokoyama, Y. *Chem. Rev.* no. 100:1717-1740.

Yun, C., J. You, J. Kim, J. Huh, and E. Kim. 2009. *J. Photochem. Photobiol., C* no. 10:111-129.

Zheng, G., X. Ma, B. Liu, Y. Dong, M. Wang, and 356 Adv. Synth. Catal. 2014, 743-748. 2014. *Adv. Synth. Catal.* no. 356:743-748.

Zijlstra, R. W. J., P. T. Duijnen, B. L. Feringa, T. Steffen, K. Duppen, and D. A. Wiersma. 1997. *J. Phys. Chem. A* no. 101:9828-9836.

In: Photoisomerization
Editor: Diego Sampedro

ISBN: 978-1-53615-313-2

Chapter 2

PHOTOISOMERIZATION OF POLYPYRIDYL RUTHENIUM(II) AQUO COMPLEXES

Yuki Tanahashi[1], Shunsuke Watabe[1], Eman A. Mohamed[1], Yuta Tsubonouchi[1], Zaki N. Zahran[1], Masanari Hirahara[2,*] and Masayuki Yagi[1,†]

[1]Department of Materials Science and Technology, Faculty of Engineering, Niigata University, Niigata, Japan
[2]Department of Applied Chemistry, National Defense Academy of Japan, Yokosuka, Japan

ABSTRACT

Polypyridyl ruthenium(II) complexes have been attract much attention as promising functional molecules due to unique photochemical and photophysical properties as well as redox properties. In a context, we reported photoisomerization of *distal*-$[Ru(tpy)(pynp)OH_2]^{2+}$ (*d*-**1**) (tpy = 2,2';6',2"-terpyridine, pynp = 2-(2-pyridyl)-1,8-naphthyridine) to

* Corresponding Author's Email: hirahara@nda.ac.jp.
† Corresponding Author's Email: yagi@eng.niigata-u.ac.jp.

proximal-[Ru(tpy)(pynp)OH_2]$^{2+}$ (*p*-**1**), which had not been characterized previously for polypyridyl ruthenium(II) aquo complexes. Herein the recent progress on mechanistic insights of photoisomerization of polypyridyl ruthenium(II) aquo complexes in our group are reviewed.

1. INTRODUCTION

Polypyridyl ruthenium(II) complexes have been attract much attention as promising functional molecules due to unique photochemical [1-6] and photophysical [7-8] properties related to a photoexcited triplet metal-to-ligand charge transfer (^{3}MLCT) state as well as redox properties including electrochromism [9-10], proton-coupled electron transfer [11-14] and multi-electron catalysis [15-18]. These properties are responsible for potential applications of the polypyridyl ruthenium(II) complexes to a large variety of devices including sensors [19-20], photovoltaic cells [21-23], displays [24-25], photocatalysts [26-27] and artificial photosynthesis [15-18, 28-33].

Recently we reported stoichiometric and irreversible photo-isomerization of *distal*-[Ru(tpy)(pynp)OH_2]$^{2+}$ (*d*-**1**) (tpy = 2,2';6',2"-terpyridine, pynp = 2-(2-pyridyl)-1,8-naphthyridine) to *proximal*-[Ru(tpy)(pynp)OH_2]$^{2+}$ (*p*-**1**) (Scheme 1) [34-36], which had not been characterized previously for polypyridyl ruthenium(II) aquo complexes although a variety of photochemical reactions [1-5, 14, 37-41] of the ruthenium(II) complexes have been reported. We have attempted to comprehensively understand the mechanism of the photoisomerization of polypyridyl ruthenium(II) aquo complexes and illustrate the effect of photoisomerization on physicochemical properties comparing between the isomers [42]. In this chapter, we review the recent progress on the photoisomerization of polypyridyl ruthenium(II) aquo complexes in our group. Mechanistic insights into the photoisomerization are provided based on the results of time-resolved IR and visible spectroscopy in laser flash photolysis, theoretical investigation by DFT calculations, and control studies of derivative complexes. Subsequently, controlling of characters of photoexcited states, redox properties and functions as a water oxidation catalyst of polypyridyl ruthenium(II) aquo complexes by the photoisomerization are introduced. Finally, a new strategic synthesis of dinuclear ruthenium(II) complexes are illustrated as application development of the photoisomerization.

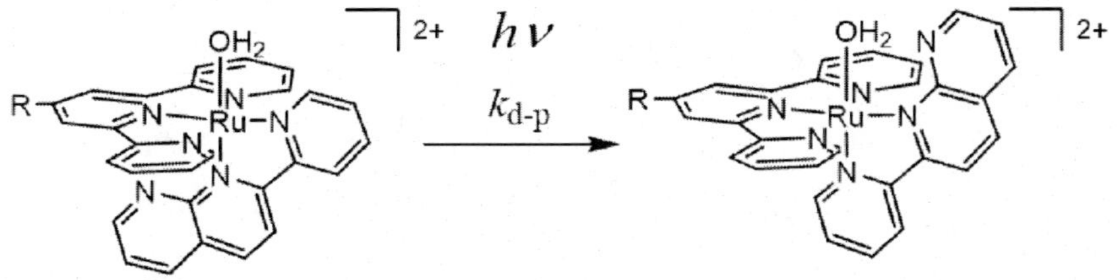

distal-$[Ru(R\text{-}tpy)(pynp)OH_2]^{2+}$ *proximal*-$[Ru(R\text{-}tpy)(pynp)OH_2]^{2+}$

R = H	*d*-**1**	*p*-**1**
R = Cl	*d*-**Cl1**	*p*-**Cl1**

Scheme 1. Photoisomerization of *distal*-$[Ru(tpy)(pynp)OH_2]^{2+}$ (*d*-**1**) to *proximal*-isomer.

2. PHOTOISOMERIZATION OF *DISTAL*-$[RU(TPY)(PYNP)OH_2]^{2+}$ TO *PROXIMAL*-ISOMER

On irradiation of visible light to a D_2O solution of *d*-**1**, the peak at 9.6 ppm assigned to *d*-**1** on the ^{1}H NMR spectrum decreased with the irradiation time and disappeared completely after 25 min under the conditions employed (Figure 1A) [34]. Alternatively, the new peak at 8.9 ppm assigned to *p*-**1** increased with the decrease of the 9.6 ppm peak. Figure 1B displays the time course of the concentration of *d*-**1** and *p*-**1**, indicating that *d*-**1** is stoichiometrically and irreversibly photoisomerized to *p*-**1** in water by visible light (Scheme 1). The kinetic profile of the photoisomerization was a first order process with respect to the *d*-**1** concentration (Figure 1B), and the kinetic analysis gave the observed rate constant ($(k_{d\text{-}p})_{obs}$ / s^{-1}) of photoisomerization (*distal* to *proximal*) to be 4.1 x 10^{-3} s^{-1} under the conditions employed ($\lambda > 420$ nm, 180 mW cm^{-2}).

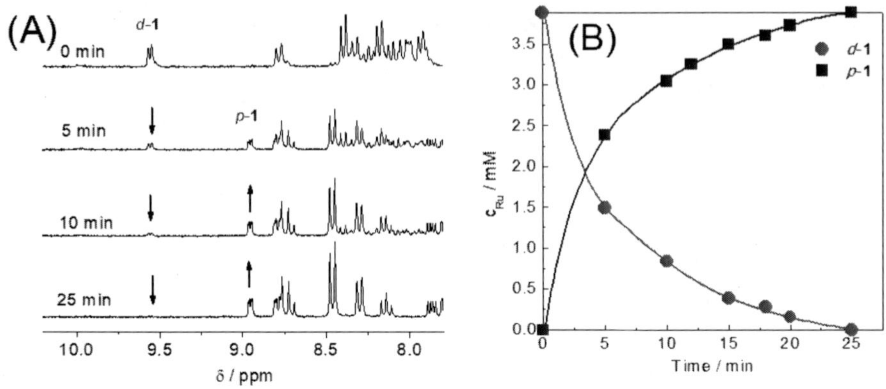

Figure 1. (A) ^{1}H NMR spectral change of *d*-**1** in D_2O with visible light irradiation ($\lambda >$ 420 nm, 180 mW cm^{-2}). (B) Plots of the concentrations (c_{Ru} / mM) of *d*-**1** (red circle) and of *p*-**1** (blue square) versus light-irradiation time. (Reproduced from ref. [34] with permission of American Chemical Society).

The internal quantum yield (Φ) for photoisomerization was provided from the UV-visible absorption spectral change in the experiment using monochromic light (520 nm, 26.4 mW cm^{-2}) and calculated according to the following eq. (1):

$$\Phi = \frac{hcN_A k_{pi} n_{int}}{p\lambda A(1-T)} \tag{1}$$

where k_{pi}, n_{int}, h, p, λ, c, N_A, T and A are the rate constant for photoisomerization, initial amount of the complex, Plank's constant, photon flux, wavelength, the speed of light, Avogadro's number, the transmittance and the irradiated area, respectively. The internal quantum yields (Φ) for the photoisomerization are 1.5 and 0.31% (λ = 520 nm) at 0.088 and 3.9 mM, respectively, as summarized in Table 1. $(k_{d\text{-}p})_{obs}$ increased linearly with the light intensity below 255 mW cm^{-2}, indicating that the photoexcited state is involved in photoisomerization under the conditions employed. The $(k_{d\text{-}p})_{obs}$ value was constant in a range of pD 1~7 (pD = -log [D$^+$]) but decreased

Table 1. Summary of the observed rate constants and internal quantum yields of photoisomerization of *d*-/*p*-1 and *d*-/*p*-2 isomers at 25°C

Complexes	λ_{max} / nm (ε / $M^{-1}cm^{-1}$)	pK_a	c_{Ru} / mM	Photoisomerization			
				$(k_{d\text{-}p})_{obs}$ / s^{-1} [a]	$(k_{p\text{-}d})_{obs}$ / s^{-1} [a]	Φ % (λ = 520 nm)	E_a / kJ mol^{-1}
***d*-1**	527 (9,300)	9.7	0.088	n.m.	-	1.5	n.m.
			3.9	4.1 x 10^{-3}	-	0.31	41.7
***p*-1**	524 (9,300)	10.7	3.9	-	n.r	-	-
***d*-Cl1**	524 (8,600)	9.3	0.088	n.m.	-	2.1	n.m.
			3.9	4.9 x 10^{-3}	-	n.m.	
***p*-Cl1**	518 (8,900)	10.9	3.9	-	n.r		
***d*-2**	501 (8,300)	9.4	3.9	1.2 x 10^{-2}	-	1.1	23.3
***p*-2**	502 (8,700)	10.5	3.9	-	4.3 x 10^{-3}	0.34	30.6

c_{Ru} is concentrations of ruthenium complexes. The marks of "n.m." and "n.r." mean "not measured" and "no reaction," respectively.

[a] The filtered halogen lamp was used for visible light irradiation (λ > 420 nm, 180 mW cm^{-2}).

drastically at pD > 7. The UV-visible absorption spectral change of the *d*-**1** solution in pH titration gave pK_a = 9.7 for deprotonation of an aquo ligand of *d*-**1** to form the hydroxo complex, *distal*-[Ru(tpy)(pynp)OH]$^+$. The pH dependency of $(k_{d\text{-}p})_{obs}$ corresponds to the distribution fraction of *d*-**1** (aquo form) versus pH, suggesting that the deprotonation of the aquo ligand of *d*-**1** makes it inert for the photoisomerization [34-35]. The photoisomerization is also temperature-dependent, and Arrhenius plots gave a straight line in a range 10 ~ 35°C to provide 41.7 kJ mol^{-1} of activation energy (E_a) for the photoisomerization (Table 1) [34-35].

3. Mechanism of Photoisomerization of *Distal*-[Ru(tpy)(pynp)OH$_2$]$^{2+}$

A possible mechanism for the photoisomerization was speculated based on the previous report on photo-substitution reactions *via* the triplet metal centered (^{3}MC) state from the excited ^{3}MLCT state for various polypyridyl complexes [1-3]. The excited ^{3}MLCT state *d*-1 is generated by a visible light irradiation, and subsequently the photo-dissociation of the aquo ligand from *d*-1 could occur *via* the thermal accessible ^{3}MC state to form a penta-coordinated [Ru(tpy)(pynp)]$^{2+}$ intermediate. The *p*-1 isomer is supposed to be formed by re-coordination of a water molecule to the intermediate from the opposite side versus a tpy plane. The temperature dependence of the transient visible absorption spectral change of the *d*-1 solution by a nanosecond laser photolysis suggests existence of the thermally activated process from the ^{3}MLCT state with the activation energy (49 kJ mol^{-1}) [35]. The transition from the ^{3}MLCT state to the ^{3}MC state could be responsible for the thermally activated process, because this value is close to the activation energy from the ^{3}MLCT state to the ^{3}MC state reported for a variety of polypyridyl ruthenium complexes [43-46]. The agreement of this activation energy (49 kJ mol^{-1}) with that (41.7 kJ mol^{-1}) for the photoisomerization reaction suggests that the thermally activated process from the ^{3}MLCT state to the ^{3}MC state might be a main activation process of the photoisomerization reaction. However, Density functional theory (DFT) calculations suggested another activation process for the conformation change of the *distal*-penta-coordinated intermediate maintaining the otherwise octahedral structure of *d*-1 to *proximal*-penta-coordinated intermediate for *p*-1, the latter of which is lower in energy than the former in the singlet and triplet spin states, as follows [35].

DFT calculations were carried out in order to elucidate the mechanism of photoisomerization of *d*-1 (Figure 2). The proposed mechanism for the photoisomerization starts with the photoexcitation of *d*-1 (closed shell singlet, *d*–S_0) to a first singlet excited state (*d*-S_1). Intersystem crossing to a first triplet excited state (*d*-T_1) state that is 33.0 kJ mol^{-1} lower in energy takes place next. Investigation of the molecular orbital changes associated with the generation of the *d*-S_1 and *d*–T_1 states from *d*–S_0 indicates that the respective excitations can be characterized as metal to ligand charge transfer (^{3}MLCT) transitions. In the Franck-Condon *d*-T_1 state, the interaction between the ruthenium center and the water molecule is found to be repulsive at all distances, so that the water molecule dissociates to form a penta-coordinated *d*–T_{penta} structure; the SOMOs of the *d*–T_{penta} structure are dominated by ruthenium d orbitals, indicating that along the water dissociation coordinate, there is an electronic state crossing that takes place leading to the pentacoordinate ^{3}MC state (Figure 2). An optimized structure for *distal*-$[Ru(tpy)(pynp)OH]^+$ with triplet spin state was also calculated, and it shows that the coordination of the hydroxyl ligand is sustained in contrast to the aquo case where the interaction between the ligand and ruthenium center is repulsive. Two competing pathways are possible following the release of the water molecule from *d*–T_1 and the generation of *d*–T_{penta}. One pathway starts with the decay of *d*–T_{penta} to ground-state *d*–S_{penta}, followed by a rate determining *d*–S_{penta} isomerization to *p*–S_{penta}, and then coordination of a water molecule to *p*–S_{penta} in order to generate *p*-1 (*p*–S_0). The free energy of activation for the isomerization step (*d*–S_{penta} → *p*–S_{penta}) is calculated to be 64.5 kJ mol^{-1}. An alternative pathway involves the pentacoordinate isomerization taking place on the triplet surface, i.e., *d*–T_{penta} to *p*–T_{penta}, followed by decay to ground state *p*–S_{penta} and subsequent water coordination to generate again *p*-1 (*p*–S_0). Unfortunately, a true transition-state structure for the isomerization step on the triplet surface (*d*–T_{penta} → *p*–T_{penta}) could not be located; however, constrained optimizations varying N-Ru-N angles to convert one isomer to another led to an approximate activation energy lower than that computed for the singlet surface. In each spin state, the *proximal* pentacoordinated structure is predicted to be lower in energy than the *distal* isomer. The actual path followed experimentally will depend on the relative rates of triplet decay and triplet isomerization; indeed, it may be that isomerization proceeds competitively on both surfaces.

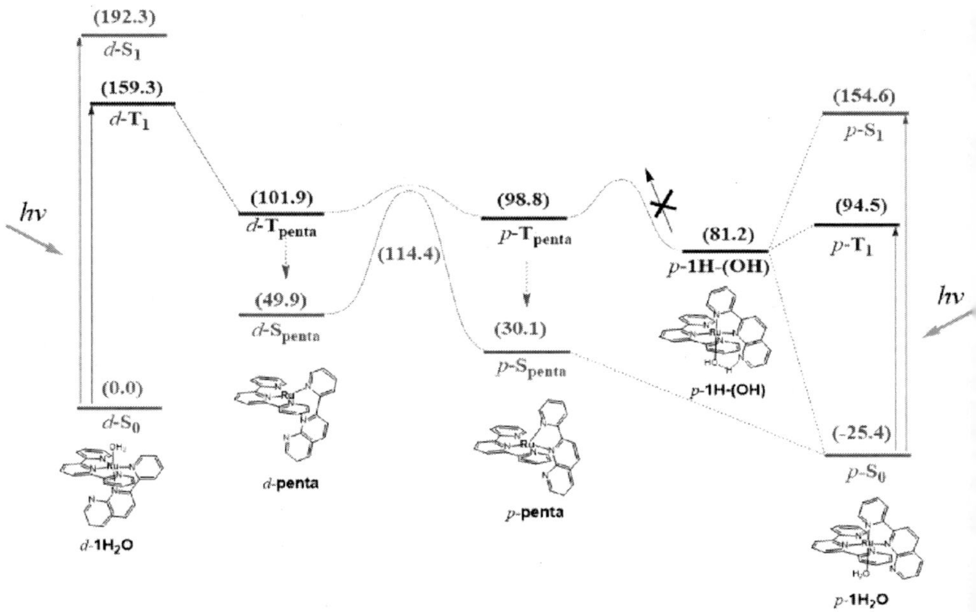

Figure 2. Energy diagram for the photoisomerization of *p*-**1** process based on DFT calculation. Red and blue curves denote the singlet and the triplet spin surfaces, respectively. All energies are calculated on the basis of that of *d*-$\mathbf{S_0}$ and reported in units of kJ mol^{-1}.

We have also investigated the possible photoisomerization of *p*-1 and find that the photoexcitation of *p*-1 (closed shell singlet, *p*–S_0) can similarly access, through an intervening S_1 state, a triplet excited-state structure having a bound H_2O molecule (named as *p*–T_1). Interestingly, *p*–T_1 is in equilibrium with the triplet excited-state of *p*-1H-(OH) which is 13.3 kJ mol^{-1} more stable and features a proton transfer from the aquo ligand to the pynp ligand so as to generate a structure having a hydroxo coordinated ruthenium center. In contrast to the triplet excited state of *d*-1, the aquo and hydroxo ligands stay bound to the ruthenium center in *p*–T_1 and *p*-1H-(OH), respectively, as they are stabilized by hydrogen bonding interactions. As a result, the photoexcited states decay directly to the ground state and back-isomerization to the *distal* form is predicted to be unfavorable.

Although the UV-visible absorption spectra of the aqueous solution is very close between *d*- and *p*-1 (only a little difference of the MLCT band) [35], only *d*-1 undergoes photoisomerization to *p*-1. (The reverse reaction was not observed.) To provide insights into the mechanism of the irreversible photoisomerization, the excited ^{3}MLCT states of the *d*- and *p*-1 isomers were characterized by time-resolved infrared (TR-IR) spectroscopic technique, which is a powerful tool to explore excited molecules because the complex vibrational spectral pattern is very sensitive to local structures and charge distributions in the molecules. The IR spectra of the photoexcited ^{3}MLCT states of both *d*- and *p*-1 isomers were characterized, and their lifetimes were given by analysis of the TR-IR spectra decays [47]. The lifetime (9.7 ns) of the excited ^{3}MLCT state of *d*-1 was longer than that (6.4 ps) of *p*-1 by three orders of magnitude. In general, excited ^{3}MLCT states of Ru polypyridyl complexes exhibit long lifetimes in the nanosecond range or above, because the transition from a triplet excited state to a singlet ground state is forbidden. The 6.4 ps lifetime observed for *p*-1 is unusually short, implying acceleration of a non-radiative process from the ^{3}MLCT state. The short lifetime of *p*-1 in the ^{3}MLCT state could be correlated to being inert for photoisomerization of *p*-1 to *d*-1.

In order to explain the dramatically different lifetimes of the ^{3}MLCT state between *d*- and *p*-1 isomers, Gibbs free energies and optimized structures of both *d*- and *p*-1 isomers in the singlet ground (S_0) and ^{3}MLCT (T_1) states were calculated by DFT. Both the structures of *d*- and *p*-1 isomers in the S_0 states (indicated by lighter color atoms in Figure 3) did not show considerable distortion. However, the aquo ligand was restricted by hydrogen bond (1.48 Å) of its H atom with an N atom on the pynp ligand for *p*-1. This is consistent with the X-ray crystallographic structures [34]. However, *d*-1 in the T_1 state had significant distortion of the pynp ligand (161° of the dihedral angle change of the pynp plane) with bending at the bond between the naphthyridine and pyridine moieties in contrast to no considerable distortion for *p*-1 in the T_1 state (Figure 3). The distortion for *d*-1 is caused by the steric hindrance between the extended π*-orbital of the pynp ligand and the π-orbital of the tpy ligand due to the charge localization on the pynp ligand in the T_1 state. Moreover, *p*-1 in the T_1 state shows the shorter Ru-O (1.98 Å) and hydrogen bonds (1.06 Å) between the pynp and the water ligands, compared with the Ru-O (2.12 Å) and hydrogen bonds (1.48 Å) in the S_0 state. The significant shortening of the hydrogen bond is also due to the charge localization on the pynp ligand on *p*-1 in the T_1 state.

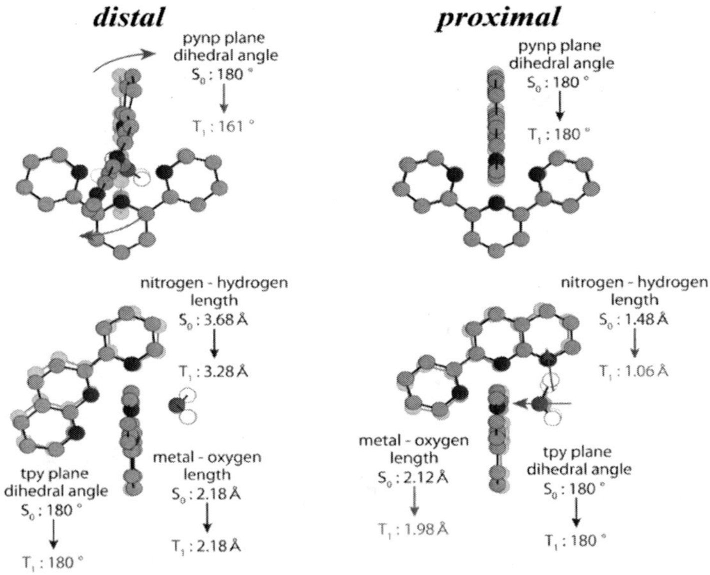

Figure 3. Optimized structures by DFT calculations of *d*- and *p*-**1** in the S_0 and T_1 states notifying characteristic changes of structural parameters accompanying the S_0 to T_1 transition. The top and bottom structures are viewed from the horizontal directions of the pynp and tpy planes, respectively. Lighter color atoms represent the S_0 state structure to make the structure change on the S_0 to T_1 transition clearer. The colored labels of atoms are follows: green, ruthenium; gray, carbon; blue, nitrogen; red, oxygen; white, hydrogen. (Reprinted with permission from ref. [47] Copyright 2015 Elsevier).

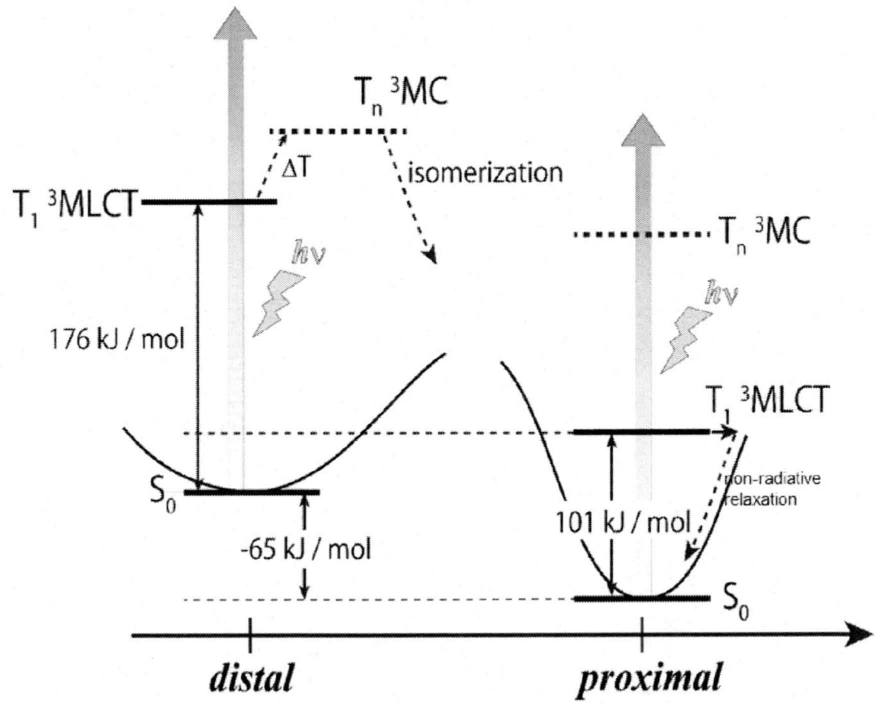

Figure 4. Schematic diagram of calculated energies of *d*- and *p*-**1** in the S_0 and T_1 states. The speculated energy levels of T_n ^{3}MC states were included. (see text) (Reprinted with permission from ref. [47] Copyright 2015 Elsevier).

Figure 4 illustrates the calculated energy levels of *d*- and *p*-1 in the S_0 and T_1 states including those in putative ^{3}MC states. *p*-1 in the S_0 states is more stable compared with *d*-1 by 65 kJ mol^{-1} due to hydrogen bond interaction between the pynp and the aquo ligand. The energy difference (176 kJ mol^{-1}) between the S_0 and T_1 states for *d*-1 is significantly higher than that (101 kJ mol^{-1}) for *p*-1. The higher energy difference for *d*-1 is attributed mainly to the significant distortion of the pynp ligand on the transition from the S_0 state to the T_1 state (Figure 3). Consequently, the energy of *p*-1 in the T_1 state is lower than that of *d*-1 in the T_1 state by 140 kJ mol^{-1}. While, the energy difference in the ^{3}MC state between *d*- and *p*-1 is assumed to be not as much as that (140 kJ mol^{-1}) in T_1 states based on the similar ligand field for both isomers. Accordingly, the energy level in the ^{3}MC state is supposed to be considerably higher than that in the T_1 state for *p*-1. This suggests the possibility of the different decay mechanism of *p*-1 in the T_1 state from the case of *d*-1 (usual lifetimes of 9.7 ns) exhibiting the non-radiative decay *via* the thermally populated ^{3}MC state. The possible mechanism is direct relaxation of the T_1 state to the S_0 state according to the energy-gap law: the decay rate decreases exponentially with increasing the energy gap between excited and ground states [48-50]. Taking almost the same structure geometries of *p*-1 between the S_0 and T_1 states into account, the law can be applied as the weak coupling limit. The lower energy band gap of 101 kJ mol^{-1} between the S_0 and T_1 states for *p*-1 could explain the short lifetime at least qualitatively according to the energy gap law without the non-radiative decay *via* the ^{3}MC state. The direct relaxation mechanism is consistent with the non-thermal decay process suggested by the very quick decay (unusual lifetime of 6.4 ps) for *p*-1 in the T_1 state. The direct relaxation process from the T_1 states to the S_0 state (not *via* the ^{3}MC state) could explain the inert photoisomerization of *p*-1.

4. Influence of Chloro Substituent on Photoisomerization

distal-[Ru(Cl-tpy)(pynp)OH_2]$^{2+}$ (*d*-Cl1) (Cl-tpy = 4'-chloro-2,2';6',2"-terpyridine) complexes were synthesized to investigate influence of chloro substituent on photoisomerizaotion of *d*-1. When an aqueous solution of *d*-Cl1 was irradiated with visible light (λ > 420 nm, 180 mWcm^{-2}), the photoisomerization of *d*-Cl1 to *proximal*-[Ru(Cl-tpy)(pynp)OH_2]$^{2+}$ (*p*-Cl1) was observed as it is for *d*-1 (Scheme 1). The rate constant for photoisomerization of *d*-Cl1 was given to be $(k_{d\text{-}p})_{obs}$ = 4.9 x 10^{-3} s^{-1} [51], which is higher than that ($(k_{d\text{-}p})_{obs}$ = 4.1 x 10^{-3} s^{-1}) [35, 52] observed for *d*-1 under the same conditions. The Φ value of *d*-Cl1 was given to be 2.1%, which is higher than that (Φ =1.5%) of *d*-1. According to the energetic scheme in Figure 4, the energy difference of *d*-Cl1 between the T_1 and ^{3}MC states could be lower than that of *p*-Cl1.

5. Reversible Photoisomerization between *Distal-* and *Proximal-*$[Ru(tpy)(pyqu)OH_2]^{2+}$

The photoisomerization between *distal-* and *proximal*-[Ru(tpy)(pyqu)$OH_2]^{2+}$ (*d*- and *p*-2) isomers with a ligand of 2-(2-pyridyl)quinoline (pyqu) instead of pynp was examined to reveal the influence of the pynp ligand on the photoisomerization of *d*-1 (See Scheme 2) [53]. When *p*-2 in D_2O was irradiated with visible light, the peak at 8.9 ppm assigned to *p*-2 in the ^{1}H NMR spectrum decreased with an increase of a doublet at 9.5 ppm assigned to the *d*-2 isomer (Figure 5A). The ^{1}H NMR spectral change was saturated in 5 min, but the peak at 8.9 ppm did not disappear completely, suggesting that the photoisomerization of *p*-2 to *d*-2 reach a photostationary state in D_2O. The equilibrated concentration ratio of *p*-2: *d*-2 was calculated to be 76:24 from the integrated peak areas at 8.9 and 9.5 ppm (Figure 5B). The kinetic profiles of the photoisomerization of *p*-2 to *d*-2 were analyzed using a reversible reaction model, and the best fit was obtained when the observed rate constants of the forward reaction (*proximal* to *distal*, $(k_{p\text{-}d})_{obs}$ / s^{-1}) and the back reaction (*distal* to *proximal*, $(k_{d\text{-}p})_{obs}$ / s^{-1}) are taken to be $(k_{p\text{-}d})_{obs} = 4.3 \pm 0.1\times 10^{-3}\ s^{-1}$ and $(k_{d\text{-}p})_{obs} = 1.2 \pm 0.03 \times 10^{-2}\ s^{-1}$ under the conditions employed ($\lambda > 420$ nm, 180 mW cm^{-2}), respectively (Table 1) [53]. The photoisomerization equilibrium between *p*- and *d*-2 isomers (Scheme 2) thus occurs in contrast to the above-mentioned *irreversible* photoisomerization from *d*-1 to *p*-1 (Scheme 1) [34-35]. The observed equilibrium constant were given to K_{obs} (= $(k_{p\text{-}d})_{obs}$ / $(k_{d\text{-}p})_{obs}$) = 0.34 ± 0.01 under the conditions employed. Both $(k_{p\text{-}d})_{obs}$ and $(k_{d\text{-}p})_{obs}$ increased linearly with light intensity up to 180 mW cm^{-2}, suggesting that the photoexcited states of *p*- and *d*-2 isomers are also involved for the forward and back reactions, respectively [53]. $(k_{p\text{-}d})_{obs}$ and $(k_{d\text{-}p})_{obs}$ deceased with pD above pD = 8, and the photoisomerization did not proceed at all above pD = 12, as shown in Figure 6 [53]. This result suggests that both hydroxo isomers, *proximal*- and *distal*-$[Ru(tpy)(pyqu)OH]^+$ are inert for the forward and backward photoisomerization reactions, respectively, as is the case for *distal*-$[Ru(tpy)(pynp)OH]^+$ [34-35]. The analysis of pD-dependent $(k_{p\text{-}d})_{obs}$

and $(k_{d\text{-}p})_{obs}$ showed the inflection points at pD = 10.5 and 9.8 for the forward and back reactions, respectively. These are close to the pK_a values of *p*-2 (pK_a = 10.5) and *d*-2 (pK_a = 9.4) given from the ^{1}H-NMR chemical shifts of protons sensitive to deprotonation of the aquo ligands, respectively (Table 1). The difference in pK_a values (1.1) between *p*-and *d*-2 results in a markedly pD-dependent K_{obs} as shown in the inset of Figure 6; K_{obs} increases above pH = 9, reaching its maximum value (K_{obs} = 2.5) at pH = 12. This is consistent with the observation that the yield of *d*-2 generated in the photoisomerization increased from 26% to 65% on increasing from pH 5.7 to 12.

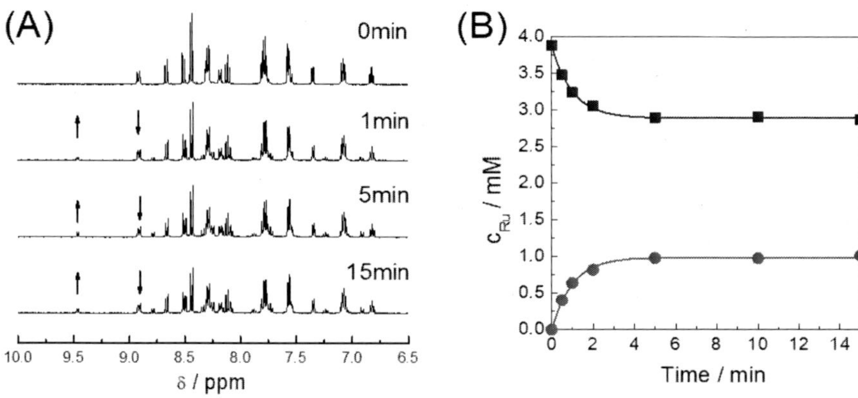

Figure 5. (A) ^{1}H NMR spectral change of *p*-**2** (3.9 mM, pD = 8.4) during light irradiation (λ > 420 nm, 180 mW/cm^2) in D_2O at 25°C. (B) Kinetic profiles of *p*-**2** (8.94 ppm) (blue square) and *d*-**2** (9.46 ppm) (red circles). (Reproduced from ref. [53] with permission of John Wiley & Sons, Inc.).

The temperature dependences of $(k_{p\text{-}d})_{obs}$ and $(k_{d\text{-}p})_{obs}$ in a range of 25 ~ 70°C provided activation energy (E_a) = 30.6 ± 2.9 and 23.3 ± 2.1 kJ mol^{-1} for the forward and backward reactions, respectively. The higher E_a for the forward reaction ($(k_{p\text{-}d})_{obs}$) compared with the backward reaction ($(k_{d\text{-}p})_{obs}$) is attributed to the unfavorable loss of the hydrogen bond (O···H-C) between the O atom of the aquo ligand and the H atom on 8-C of a quinoline moiety of pyqu upon water dissociation for photoisomerization. The van't Hoff plots for K_{obs} in the same temperature range gave ΔH = 7.7 ± 2.7 kJ mol^{-1}, which is close to the hydrogen bond enthalpy (5.4 kJ mol^{-1}) of O···H-C in the water-benzene complex [54]. The van't Hoff plot shows that the forward photoisomerization of *p*-2 to *d*-2 proceeds endothermically, which is consistent with the higher yield (35%) of *d*-2 in the photoisomerization at 70°C than that (26%) at 25°C under the neutral pD conditions [53].

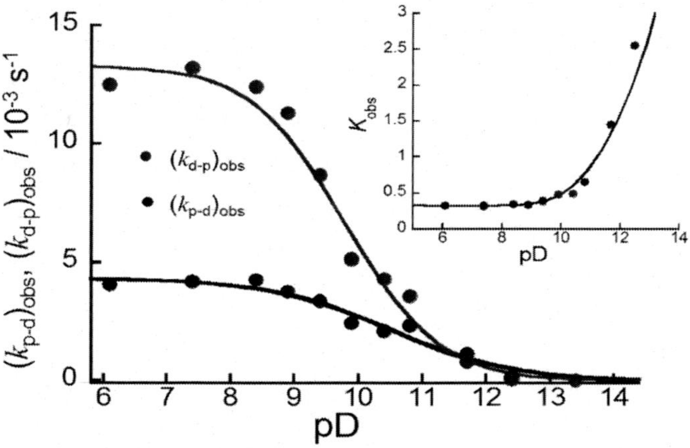

Figure 6. Plots of observed rate constants (($k_{p\text{-}d})_{obs}$ and $(k_{d\text{-}p})_{obs}$) vs pD for the photoisomerization reactions shown in Scheme 2. Visible light (λ > 420 nm, 180 mW/cm^2) was irradiated to the *p*-**2** solution (3.9 mM) in D_2O at 25°C. Inset shows plots of observed equilibrium constants K_{obs} vs pD. (Reproduced from ref. [53] with permission of John Wiley & Sons, Inc.).

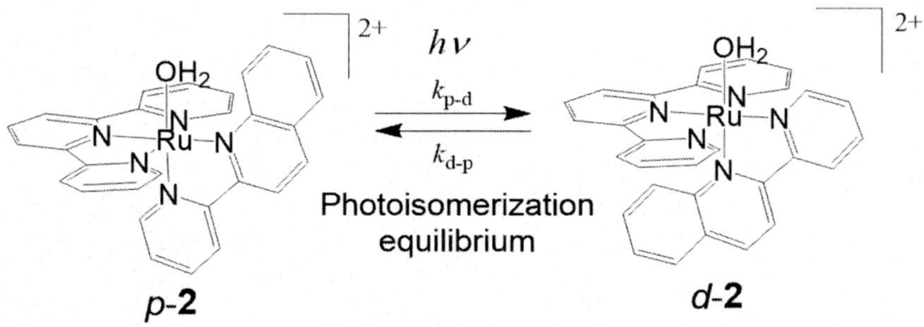

Scheme 2. Photoisomerization equilibrium of *proximal*- and *distal*-$[Ru(tpy)(pyqu)OH_2]^{2+}$ (*p*- and *d*-**2**) isomers.

Φ for the forward and back photoisomerization reactions were assessed to be 0.34% and 1.1% (3.9 mM) from the experiments using monochromatic light (520 nm), as summarized in Table 1. The higher Φ (1.1%) for *d*-2 than that (0.34%) for *p*-2 could be explained by a combination of contributions of the hydrogen bond (O···H-C) enthalpy for *p*-2 (lowered Φ for *p*-2) and the dominant steric repulsion between the H atom on 8-C of the quinoline moiety of pyqu and the tpy plane for *d*-2 (enhanced Φ for *d*-2). The Φ value (1.1%) for photoisomerization from *d*-2 to *p*-2 is 3.5 times higher than that (0.31%) for the photoisomerization from *d*-1 to *p*-1 under the same concentration [53]. This could be also ascribable to the dominant steric repulsion between the H atom on 8-C of the quinoline moiety of pyqu and the tpy ligand for *d*-2, compared with that between the uncoordinated nitrogen on the naphthyridine moiety of pynp and the tpy ligand for *d*-1.

6. Mechanism of Reversible Photoisomerization between *Distal-* and *Proximal-*[Ru(tpy)(pyqu)OH$_2$]$^{2+}$

DFT calculations were carried out to understand the reversible photoisomerization between *p*- and *d*-2 [53]. Upon absorption of a photon, *d*-2 undergoes a transition to a singlet excited state (*d*-S$_1$) followed by spin-orbit facilitated intersystem crossing to the triplet surface (Figure 7). On the triplet surface, the ruthenium-aquo interaction becomes much less attractive, and release of bound water to form a penta-coordinate intermediate (*d*-Ru-penta) maintaining the otherwise octahedral structure of *d*-2 (9.2 kJ mol^{-1}) is less endergonic by 36.1 kJ mol^{-1} compared to the same process on the ground-state potential energy surface (45.3 kJ mol^{-1}). *d*-Ru-penta then converts to *p*-Ru-penta, i.e., a penta-coordinate intermediate reflecting the otherwise octahedral structure of *p*-RuH$_2$O. We calculated the energy of the relevant transition-state structures (TS-penta) on both the triplet and singlet surfaces. The barrier for the transition (4.9 kJ mol^{-1}) from *d*-Ru-penta to *p*-Ru-penta on the triplet surface is lowered by 50.2 kJ mol^{-1} relative to that (55.1 kJ mol^{-1}) on the singlet ground-state surface. These energetics overall suggest that the conversion of *d*-Ru-penta to *p*-Ru-penta occurs predominantly on the triplet surface. The *p*-Ru-penta structure can then relax to its ground-state singlet analog, to which coordination of a water molecule is thermodynamically favorable. The inverse *proximal* to *distal* photoisomerization is predicted to occur in much the same way, with the minor difference that upon relaxation from the excited-state singlet to the triplet state, *p*-2 dissociates coordinated water through a barrierless process to become *p*-Ru-penta, i.e., the triplet excited state of *p*-2 is predicted to be unstable as a hexa-coordinate structure. This is consistent with the significant steric repulsion between the aquo ligand and the H atom on 8-C of a quinoline moiety. The lower barrier for conversion of *p*-Ru-penta to *d*-Ru-penta on the triplet surface (8.9 kJ mol^{-1}) compared to the singlet (net 115.9 kJ mol^{-1}) suggests that the conversion of *p*-Ru-penta to *d*-Ru-penta also occurs for the former spin state. Finally, as determined from the experimental van't Hoff plot, the isomerization of *p*-2 to *d*-2 is predicted to be endothermic.

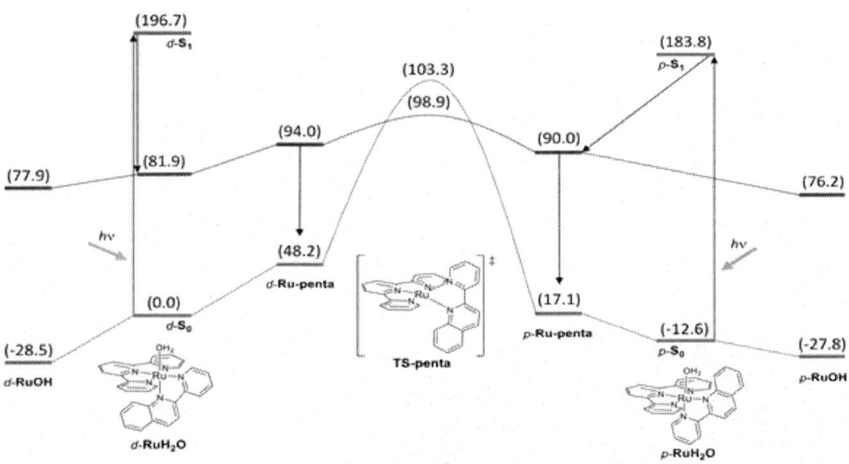

Figure 7. Energy diagram for the photoisomerization equilibrium of *p-/d-***2**. Curves in blue and red represent the singlet and the triplet spin surfaces, respectively. All energies are calculated on the basis of that of *d*-$\mathbf{S_0}$ (at pH = 14) and reported in units of kJ mol^{-1}. (Reprinted with permission from ref. [53] Copyright 2015 John Wiley & Sons, Inc.).

7. CONTROLS OF PROPERTIES AND FUNCTIONS BY PHOTOISOMERIZATION

7.1. Control of Redox Properties

For the cyclic voltammogram (CV) of *d*-**1** in an aqueous solution, two redox waves were given at 0.46 V and 0.63 V vs the saturated calomel electrode (SCE) at pH = 6.4 (Figure 8A) [34-35]. These potentials are pH-dependent in a wide range of pH = 2 ~ 10 (54 mV / pH for both the redox waves), indicating a 2-step reaction of 1-proton-coupled 1-electron reactions of Ru^{II}-OH_2/Ru^{III}-OH and Ru^{III}-OH / Ru^{IV}=O redox couples for *d*-**1** in the pH range, as shown by Pourbaix diagrams (Figure 8B) [34]. At pH = 10 ~ 13, a 2-step reaction of a (non-proton coupled) 1-electron reaction of Ru^{II}-OH / Ru^{III}-OH followed by a 1-proton-coupled 1-electron reaction of Ru^{III}-OH / Ru^{IV}=O. Under acidic conditions (below pH = 2), Ru^{IV}=O is formed by a 2-step reaction of a (non-proton coupled) 1-electron reaction of a Ru^{II}-OH_2 / Ru^{III}-OH_2 redox couple followed by a 2-proton-coupled 1-electron reaction of a Ru^{III}-OH_2 / Ru^{IV}=O redox couple. When *d*-**1** underwent photoisomerization to *p*-**1** by visible light irradiation, CV was significantly changed; CV of *p*-**1** (after irradiation) gave only a redox wave at 0.81 V vs SCE at pH = 6.4 (Figure 8A) [34-35]. This suggests that a 1-step reaction of a 2-proton-coupled 2-electron reaction of a Ru^{II}-OH_2 / Ru^{IV}=O redox couple occurs at pH = 6.4, being consistent with Pourbaix diagrams at pH = 2 ~ 10 (59 mV / pH) (Figure 8A). This illustrates the redox properties controlled by photoisomerization of ruthenium(II) aquo complexes. The CV data and Pourbaix diagrams of *d*-/*p*-**Cl1** [51] and *d*-/*p*-**2** [53] in an aqueous solution showed that similar proton-coupled electron transfer (PCET) reactions to *d*- and *p*-**1** occurs, suggesting that influences of chloro substitution to pynp and alternation to the pyqu ligand on the PCET reaction mechanism of the ruthenium aquo complexes are not considerably observed.

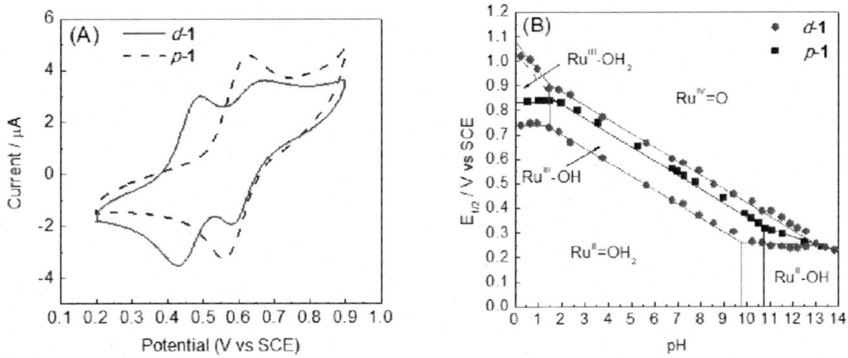

Figure 8. (A) Cyclic voltammograms of 0.5 mM *d*-**1** (red line) and *p*-**1** (dashed blue line) in a 0.1 M phosphate buffer solution (pH = 6.4). Scan rate: 20 mV s^{-1}. A grassy carbon electrode was used as a working electrode. (Reprinted with permission from ref. [35] Copyright 2013 American chemical Society.) (B) Pourbaix diagrams for *d*-**1** (red circle) and *p*-**1** (blue square) in an aqueous solution. $E_{1/2}$ is represented as a half wave potential and calculated from the midpoint of anodic and cathodic peak potentials. (Reproduced from ref. [34] with permission of American chemical Society).

7.2. Control of Catalytic Activities for Water Oxidation

Water oxidation catalysts have attracted much attention to construct artificial photosynthesis that is expected as chemical conversion system of solar energy [28, 32, 55-59]. The catalytic activity of *d*-1 for water oxidation was examined in an aqueous solution using ammonium cerium(IV) nitrate as an oxidant. O_2 was evolved when *d*-1 and the oxidant were mixed in the solution, showing that *d*-1 is active for water oxidation catalysis (Figure 9) [34-35]. However, the amount of O_2 evolved is significantly decreased after *d*-1 underwent photoisomerization to *p*-1 by visible light irradiation. The activity analysis indicates that the turnover frequency (TOF = 4.8 x 10^{-4} s^{-1}) of *p*-1 is lower than that (3.8 x 10^{-3} s^{-1}) of *d*-1 by nearly an order of magnitude [34-35]. This result shows that the observed catalytic activity for water oxidation is controlled by photoisomerization of *d*-1 to *p*-1. It also tells us that we should pay attention to the use of the *p*-1 catalyst in photochemical systems such artificial photosynthesis because the observed catalytic activities are decreased by visible light irradiation.

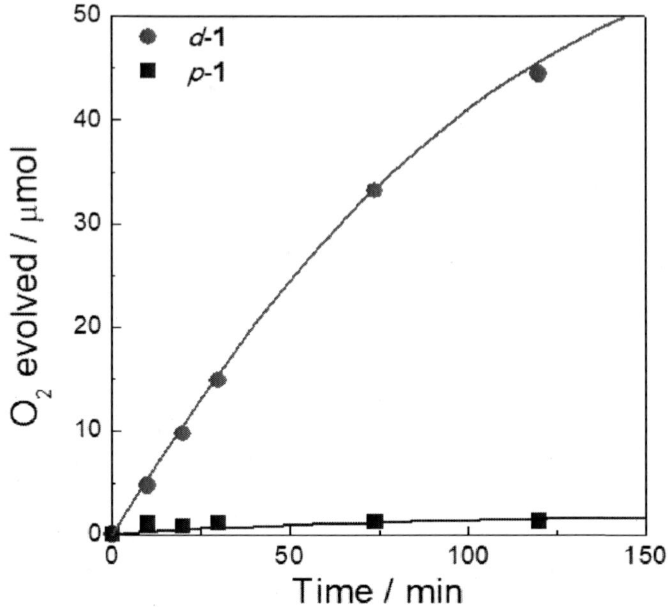

Figure 9. Time courses of the amount of O_2 evolved in an aqueous solution at 25°C during chemical water oxidation using ammonium cerium(IV) nitrate oxidant for *d*-**1** (red circle) and *p*-**1** (blue square). oxidant, 0.1 M (0.5 mmol); Ru catalysts, 1.0 μmol; pH = 1.0; liquid volume, 5.0 ml. (Reproduced from ref. [34] with permission of American chemical Society).

The CV data suggest that both isomers are oxidized to Ru^{IV}=O species which could be further oxidized to Ru^{V}=O for water oxidation [35]. The nucleophilic attack of a water molecule on Ru^{V}=O species could be presumed to form an O-O bond during the catalysis by both isomers. The DFT calculations show that water nucleophilic attack on *d*-Ru^{V}=O (derived from *d*-1, calculated $\Delta G^{\ddagger}$ of 87.9 kJ mol^{-1}) is favored over *p*-Ru^{V}=O (derived from *p*-1, calculated $\Delta G^{\ddagger}$ of 104.6 kJ/mol) for O-O bond formation [35]. Examination of the lowest unoccupied molecular orbitals in *d*- and *p*-Ru^{V}=O indicates that more orbital amplitude is concentrated on the [Ru-O] unit in the case of *d*-Ru^{V}=O than in the case of the *proximal*-isomer, where some of the amplitude is instead delocalized over the pynp ligand, making this isomer less electrophilic (Figure 10) [35].

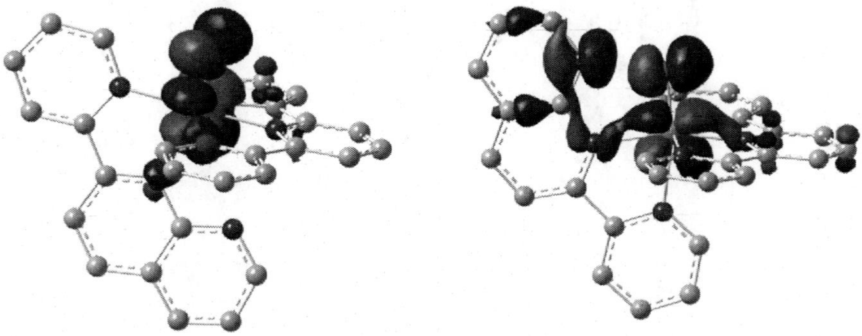

Figure 10. Molecular orbitals of the lowest unoccupied molecular orbitals (LUMO 1 (127 α)) of *distal*-[Ru^{V}(tpy)(pynp)(O)]$^{3+}$ (*d*-Ru^{V}=O, left) and *proximal*-[Ru^{V}(tpy)(pynp)(O)]$^{3+}$ (*p*-Ru^{V}=O, right). Carbon atoms are gray, nitrogen atoms blue, oxygen atoms red, and Ru atoms cyan; hydrogen atoms are not shown for clarity. (Reprinted with permission from ref. [35] Copyright 2013 American chemical Society).

8. STRATEGIC SYNTHESIS OF DINUCLEAR RUTHENIUM(II) COMPLEXES USING PHOTOISOMERIZATION

The photoisomerization of ruthenium(II) aqua complexes was extended to strategic synthesis of dinuclear ruthenium complexes that is difficult to be synthesized by conventional thermal reactions so far. Herein we show unique synthesis of new series of dinuclear ruthenium(II) complexes, *proximal,proximal*-[$Ru_2(tpy)_2LXY$]$^{3+}$ (L = 5-phenyl-2,8-di(2-pyridyl)-1,9,10-anthyridine, X and Y = the other coordination sites; denoted as *p,p*-Ru_2XY) using the photoisomerization of a mononuclear complex (See Scheme 3) [60]. A multidentate bridging ligand L was utilized to construct a *p,p*-dinuclear ruthenium(II) structure. The reaction of [Ru(tpy)Cl_3] and L in ethanol selectively gave *distal*-[Ru(tpy)LCl]$^{+}$ (*d*-3Cl, 59% yield). The selective formation of *d*-3Cl is possibly due to steric hindrance of L with a chloro ligand. *d*-3Cl was easily converted to *distal*-[Ru(tpy)LOH_2]$^{2+}$ (*d*-3) by chloro subtraction with $AgNO_3$ in water (90% yield). The thermal reaction of *d*-3 with [Ru(tpy)Cl_3] as a second ruthenium center for dimerization in water failed to yield *p,p*-dinuclear ruthenium(II) species due to the steric hindrance between a tpy ligand on *d*-3 and one on [Ru(tpy)Cl_3]. However, if photoisomerization of *d*-3 to *proximal*-[Ru(tpy)LOH_2]$^{2+}$ (*p*-3) can be employed as shown in Scheme 3, the steric constraint of *d*-3 for formation of the *p,p*-dinuclear ruthenium(II) species could be sidestepped. The photoisomerization of *d*-3 stoichiometrically proceeded to form *p*-3 in water/ethanol (4:1 vol/vol) under visible light irradiation ($\lambda > 420$ nm). The subsequent thermal reaction of *p*-3 with [Ru(tpy)Cl_3] in water/ethanol (1:3 vol/vol) successfully produced *proximal,proximal*-[Ru_2(tpy)$_2$L(μ-Cl)]$^{3+}$ (*p,p*-Ru_2(μ-Cl), 67% yield), which was characterized by X-ray crystallography [60].

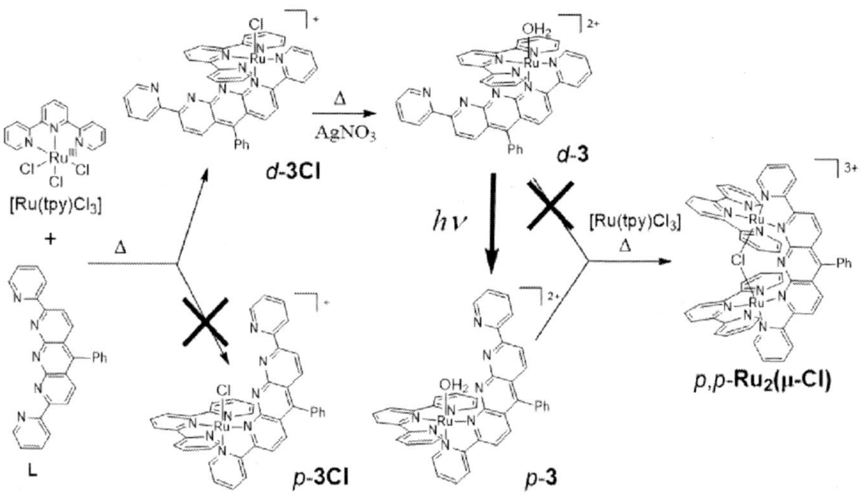

Scheme 3. Synthetic scheme of a dinuclear ruthenium complex, *p,p*-**Ru_2(μ-Cl)** using photoisomerization of *d*-**3**. (Reproduced from ref. [60] with permission of American Chemical Society).

CONCLUSION

Recent progress on the aspects and mechanistic insights of photoisomerization of mononuclear ruthenium(II) aquo complexes in our group were reviewed to understand the photoisomerization reactions and its mechanism comprehensively. The controls of properties and functions of mononuclear ruthenium(II) aquo complexes by the photoisomerization were exemplified in terms of their proton-coupled electron transfer reactions and water oxidation catalyses. An example of application of the photoisomerization is demonstrated; the employment of the photoisomerization enabled to synthesize the dinuclear ruthenium(II) complexes that have been difficult to be synthesized by conventional thermochemical processes. New design and development of different varieties of ruthenium(II) complexes are desired to explore unique reactions, functions and application based on the photoisomerization for future molecular devices.

ACKNOWLEDGMENT

This work was supported by the JST PRESTO program, JSPS KAKENHI Grant Number JP24107003, JP17H06439 and JP18H02071 in Scientific Research on Innovative Areas "All Nippon Artificial Photosynthesis Project for Living Earth (AnApple)" and "Innovations for Light-Energy Conversion (I^4LEC)."

REFERENCES

[1] Durham, B., Walsh, J. L., Carter, C. L. and Meyer, T. J.: Synthetic applications of ohotosubstitution reactions of poly(pyridyl) complexes of ruthenium(II). *Inorganic Chemistry,* 1980, 19 (4), 860-865.

[2] Pinnick, D. V. and Durham, B.: Photosubstitution reactions of $Ru(bpy)_2XY^{n+}$ complexes. *Inorganic Chemistry,* 1984, 23 (10), 1440-1445.

[3] Hecker, C. R., Fanwick, P. E. and Mcmillin, D. R.: Evidence for dissociative photosubstitution reactions of $[Ru(trpy)(bpy)(NCCH_3)]^{2+}$

- crystal and molecular-structure of [Ru(trpy)(bpy)(py)] $(PF_6)_2\cdot(CH_3)_2CO$. *Inorganic Chemistry,* 1991, 30 (4), 659-666.

[4] Howerton, B. S., Heidary, D. K. and Glazer, E. C.: Strained ruthenium complexes are potent light-activated anticancer agents. *Journal of the American Chemical Society,* 2012, 134 (20), 8324-8327.

[5] Albani, B. A., Peña, B., Leed, N. A., De Paula, N. a. B. G., Pavani, C., Baptista, M. S., Dunbar, K. R. and Turro, C.: Marked improvement in photoinduced cell death by a new tris-heteroleptic complex with dual action: singlet oxygen sensitization and ligand dissociation. *Journal of the American Chemical Society,* 2014, 136 (49), 17095-17101.

[6] Yamazaki, H., Miyagawa, T., Shoji, A., Kajita, M. and Yagi, M. (2011): Catalytic water oxidation to produce molecular oxygen by ruthenium complexes. In: Watoson, D. B. (Ed.) *Ruthenium: Properties, Production and Applications*: pp. 399-411: New York, Nova Science Publishers.

[7] Liu, F., Concepcion, J. J., Jurss, J. W., Cardolaccia, T., Templeton, J. L. and Meyer, T. J.: Mechanisms of water oxidation from the blue dimer to photosystem II. *Inorganic Chemistry,* 2008, 47 (6), 1727-1752.

[8] Hewitt, J. T., Concepcion, J. J. and Damrauer, N. H.: Inverse kinetic isotope effect in the excited-state relaxation of a Ru(II)–aquo complex: revealing the impact of hydrogen-bond dynamics on nonradiative decay. *Journal of the American Chemical Society,* 2013, 135 (34), 12500-12503.

[9] Yagi, M. and Kaneko, M.: Charge transport and catalysis by molecules confined in polymeric materials and application to future nanodevices for energy conversion. *Advances in Polymer Science,* 2006, 199, 143-188.

[10] Narita, K., Kuwabara, T., Sone, K., Shimizu, K. and Yagi, M.: Characterization and activity analysis of catalytic water oxidation induced by hybridization of $[(OH_2)(terpy)Mn(m\text{-}O)_2Mn\ (terpy)(OH_2)]^{3+}$ and clay compounds. *Journal of Physical Chemistry B,* 2006, 110 (46), 23107-23114.

[11] Yoshida, M., Masaoka, S. and Sakai, K.: Oxygen evolution from water catalyzed by mononuclear ruthenium complexes with a triazamacrocyclic ligand in a facial fashion. *Chemistry Letters,* 2009, 38 (7), 702-703.

[12] Concepcion, J. J., Jurss, J. W., Norris, M. R., Chen, Z. F., Templeton, J. L. and Meyer, T. J.: Catalytic water oxidation by single-site ruthenium catalysts. *Inorganic Chemistry,* 2010, 49 (4), 1277-1279.

[13] Wasylenko, D. J., Ganesamoorthy, C., Henderson, M. A., Koivisto, B. D., Osthoff, H. D. and Berlinguette, C. P.: Electronic modification of the $[Ru^{II}(tpy)(bpy)(OH_2)]^{2+}$ scaffold: effects on catalytic water oxidation. *Journal of the American Chemical Society,* 2010, 132 (45), 16094-16106.

[14] Padhi, S. K., Fukuda, R., Ehara, M. and Tanaka, K.: Photoisomerization and proton-coupled electron transfer (PCET) promoted water oxidation by mononuclear cyclometalated ruthenium catalysts. *Inorganic Chemistry,* 2012, 51 (9), 5386-5392.

[15] Gersten, S. W., Samuels, G. J. and Meyer, T. J.: Catalytic oxidation of water by an oxo-bridged ruthenium dimer. *Journal of the American Chemical Society,* 1982, 104 (14), 4029-4030.

[16] Wada, T., Tsuge, K. and Tanaka, K.: Electrochemical oxidation of water to dioxygen catalyzed by the oxidized form of the bis(ruthenium-hydroxo) complex in H_2O. *Angewandte Chemie International Edition,* 2000, 39 (8), 1479-1482.

[17] Sens, C., Romero, I., Rodriguez, M., Llobet, A., Parella, T. and Benet-Buchholz, J.: A new Ru complex capable of catalytically oxidizing water to molecular dioxygen. *Journal of the American Chemical Society,* 2004, 126 (25), 7798 - 7799.

[18] Zong, R. and Thummel, R. P.: A new family of Ru complexes for water oxidation. *Journal of the American Chemical Society,* 2005, 127 (37), 12802-12803.

[19] Montalti, M., Wadhwa, S., Kim, W. Y., Kipp, R. A. and Schmehl, R. H.: Luminescent ruthenium(II) bipyridyl−phosphonic acid complexes: pH dependent photophysical behavior and quenching with divalent metal ions. *Inorganic Chemistry,* 2000, 39 (1), 76-84.

[20] Anzenbacher, P., Tyson, D. S., Jursíková, K. and Castellano, F. N.: Luminescence lifetime-based sensor for cyanide and related anions. *Journal of the American Chemical Society,* 2002, 124 (22), 6232-6233.

[21] O'regan, B. and Gratzel, M.: A low-cost, high-efficiency solar cell based on dye-sensitized colloidal TiO_2 films. *Nature,* 1991, 353 (6346), 737-740.

[22] Kalyanasundaram, K. and Gratzel, M.: Applications of functionalized transition metal complexes in photonic and optoelectronic devices. *Coordination Chemistry Reviews,* 1998, 177, 347-414.

[23] Wang, P., Zakeeruddin, S. M., Moser, J. E., Nazeeruddin, M. K., Sekiguchi, T. and Gratzel, M.: A stable quasi-solid-state dye-sensitized solar cell with an amphiphilic ruthenium sensitizer and polymer gel electrolyte. *Nat Mater,* 2003, 2 (6), 402-407.

[24] Soda, T., Kitagawa, Y., Onishi, T., Takano, Y., Shigeta, Y., Nagao, H., Yoshioka, Y. and Yamaguchi, K.: Ab initio computations of effective exchange integrals for H-H, H-He-H and Mn_2O_2 complex: comparison of broken-symmetry approaches. *Chemical Physics Letters,* 2000, 319 (3-4), 223-230.

[25] Buda, M., Kalyuzhny, G. and Bard, A. J.: Thin-film solid-state electroluminescent devices based on tris(2,2'-bipyridine) ruthenium(II) complexes. *Journal of the American Chemical Society,* 2002, 124 (21), 6090-6098.

[26] Kirch, M., Lehn, J. M. and Sauvage, J. P.: Hydrogen generation by visible light irradiation of aqueous solutions of metal complexes. An approach to the photochemical conversion and storage of solar energy. *Helvetica Chimica Acta,* 1979, 62 (4), 1345-1384.

[27] Juris, A., Balzani, V., Barigelletti, F., Campagna, S., Belser, P. and Vonzelewsky, A.: Ru(II) polypyridine complexes - photophysics, photochemistry, electrochemistry, and chemi-luminescence. *Coordination Chemistry Reviews,* 1988, 84, 85-277.

[28] Concepcion, J. J., Jurss, J. W., Templeton, J. L. and Meyer, T. J.: One site is enough. Catalytic water oxidation by $[Ru(tpy)(bpm)(OH_2)]^{2+}$ and $[Ru(tpy)(bpz)(OH_2)]^{2+}$. *Journal of the American Chemical Society,* 2008, 130 (49), 16462-16463.

[29] Masaoka, S. and Sakai, K.: Clear evidence showing the robustness of a highly active oxygen-evolving mononuclear ruthenium complex with an aqua ligand. *Chemistry Letters,* 2009, 38 (2), 182-183.

[30] Xu, Y., Fischer, A., Duan, L., Tong, L., Gabrielsson, E., Åkermark, B. and Sun, L.: Chemical and light-driven oxidation of water catalyzed by an efficient dinuclear ruthenium complex. *Angewandte Chemie International Edition,* 2010, 49 (47), 8934-8937.

[31] Wasylenko, D. J., Ganesamoorthy, C., Koivisto, B. D., Henderson, M. A. and Berlinguette, C. P.: Insight into water oxidation by mononuclear polypyridyl Ru catalysts. *Inorganic Chemistry,* 2010, 49 (5), 2202-2209.

[32] Boyer, J. L., Polyansky, D. E., Szalda, D. J., Zong, R., Thummel, R. P. and Fujita, E.: Effects of a proximal base on water oxidation and proton reduction catalyzed by geometric isomers of [Ru(tpy)(pynap)$(OH_2)]^{2+}$. *Angewandte Chemie International Edition,* 2011, 50 (52), 12600-12604.

[33] Yagi, M., Tajima, S., Komi, M. and Yamazaki, H.: Highly active and tunable catalysts for O_2 evolution from water based on mononuclear ruthenium (II) monoaquo complexes. *Dalton Transactions,* 2011, 40 (15), 3802-3804.

[34] Yamazaki, H., Hakamata, T., Komi, M. and Yagi, M.: Stoichiometric photoisomerization of mononuclear ruthenium(II) monoaquo complexes controlling redox properties and water oxidation catalysis. *Journal of the American Chemical Society,* 2011, 133 (23), 8846-8849.

[35] Hirahara, M., Ertem, M. Z., Komi, M., Yamazaki, H., Cramer, C. J. and Yagi, M.: Mechanisms of photoisomerization and water-oxidation catalysis of mononuclear ruthenium(II) monoaquo complexes. *Inorganic Chemistry,* 2013, 52 (11), 6354-6364.

[36] Takahashi, K., Hirahara, M., Sato, T. and Yagi, M. (2016): Unique synthesis, properties and functions of mono- and di-nuclear ruthenium(II) aquo complexes. In: Taylor, J. C. (Ed.) *Advances in Chemistry Research*: pp. 35-50: New York, Nova Science Publishers.

[37] Durham, B., Wilson, S. R., Hodgson, D. J. and Meyer, T. J.: Cis-trans photoisomerization in $Ru(bpy)_2(OH_2)_2^{2+}$ - crystal structure of *trans-*

$[Ru(bpy)_2(OH_2)(OH)](ClO_4)_2$. *Journal of the American Chemical Society,* 1980, 102 (2), 600-607.

[38] Porter, G. B. and Sparks, R. H.: Photoracemization of Ru $(bipyridine)_3^{2+}$. *Journal of Photochemistry,* 1980, 13 (2), 123-131.

[39] Rachford, A. A. and Rack, J. J.: Picosecond isomerization in photochromic ruthenium−dimethyl sulfoxide complexes. *Journal of the American Chemical Society,* 2006, 128 (44), 14318-14324.

[40] Miyazaki, S., Kojima, T. and Fukuzumi, S.: Photochemical and thermal isomerization of a ruthenium(II)−alloxazine complex involving an unusual coordination mode. *Journal of the American Chemical Society,* 2008, 130 (5), 1556-1557.

[41] Bonnet, S., Limburg, B., Meeldijk, J. D., Gebbink, R. J. M. K. and Killian, J. A.: Ruthenium-decorated lipid vesicles: light-induced release of $[Ru(terpy)(bpy)(OH_2)]^{2+}$ and thermal back coordination. *Journal of the American Chemical Society,* 2011, 133 (2), 252-261.

[42] Hirahara, M. and Yagi, M.: Photoisomerization of ruthenium(ii) aquo complexes: mechanistic insights and application development. *Dalton Transactions,* 2017, 46 (12), 3787-3799.

[43] Allsopp, S. R., Cox, A., Kemp, T. J. and Reed, W. J.: Inorganic photophysics in solution. Part 1.-Temperature activation of decay processes in the luminescence of tris(2,2'-bipyridine)ruthenium(II) and tris(1,10-phenanthroline)ruthenium(II) ions. *Journal of the Chemical Society, Faraday Transactions 1: Physical Chemistry in Condensed Phases,* 1978, 74, 1275-1289.

[44] Durham, B., Caspar, J. V., Nagle, J. K. and Meyer, T. J.: Photochemistry of $Ru(Bpy)_3^{2+}$. *Journal of the American Chemical Society,* 1982, 104 (18), 4803-4810.

[45] Norrby, T., Börje, A., Åkermark, B., Hammarström, L., Alsins, J., Lashgari, K., Norrestam, R., Mårtensson, J. and Stenhagen, G.: Synthesis, structure, and photophysical properties of novel ruthenium(II) carboxypyridine type complexes. *Inorganic Chemistry,* 1997, 36 (25), 5850-5858.

[46] Li, C. and Hoffman, M. Z.: Electron localization or delocalization in the MLCT excited states of $Ru(bpy)_3^{2+}$ and $Ru(phen)_3^{2+}$.

Consequences to their photochemistry and photophysics in fluid solution. *Inorganic Chemistry,* 1998, 37 (4), 830-832.

[47] Tanaka, S., Takahashi, K., Hirahara, M., Yagi, M. and Onda, K.: Characterization of the excited states of *distal-* and *proximal-* $[Ru(tpy)(pynp)OH_2]^{2+}$ in aqueous solution using time-resolved infrared spectroscopy. *Journal of Photochemistry and Photobiology A: Chemistry,* 2015, 313, 87-98.

[48] Caspar, J. V., Kober, E. M., Sullivan, B. P. and Meyer, T. J.: Application of the energy gap law to the decay of charge-transfer excited states. *Journal of the American Chemical Society,* 1982, 104 (2), 630-632.

[49] Caspar, J. V. and Meyer, T. J.: Application of the energy gap law to nonradiative, excited-state decay. *The Journal of Physical Chemistry,* 1983, 87 (6), 952-957.

[50] Kober, E. M., Caspar, J. V., Lumpkin, R. S. and Meyer, T. J.: Application of the energy gap law to excited-state decay of osmium(II)-polypyridine complexes: calculation of relative nonradiative decay rates from emission spectral profiles. *The Journal of Physical Chemistry,* 1986, 90 (16), 3722-3734.

[51] Takahashi, K., Zhang, X., Hirahara, M., Sato, T., Saito, K., Yui, T. and Yagi, M.: Influence of chloro substituent on photoisomerization, redox reactions and water oxidation catalysis of mononuclear ruthenium complexes. *Journal of Photochemistry and Photobiology A: Chemistry,* 2015, 313, 117-125.

[52] Yamazaki, H., Hakamata, T., Komi, M. and Yagi, M.: Stoichiometric photoisomerization of mononuclear ruthenium(II) monoaquo complexes controlling redox properties and water oxidation catalysis. *Journal of the American Chemical Society,* 2011, 133 (23), 8846-8849.

[53] Hirahara, M., Tomoya, H., League, A. B., Ertem, M. Z., Takahashi, K., Nagai, S., Inaba, K., Yamazaki, H., Saito, K., Yui, T., Cramer, C. J. and Yagi, M.: Mechanisms and factors controlling photoisomerization equilibrium, ligand exchange and water oxidation catalysis of mononuclear ruthenium(II) complexes *European Journal of Inorganic Chemistry,* 2015, 2015 (23), 3892-3903.

[54] Li, S., Cooper, V. R., Thonhauser, T., Puzder, A. and Langreth, D. C.: A density functional theory study of the benzene−water complex. *Journal of Physical Chemistry A,* 2008, 112 (38), 9031-9036.

[55] Masllorens, E., Rodriguez, M., Romero, I., Roglans, A., Parella, T., Benet-Buchholz, J., Poyatos, M. and Llobet, A.: Can the disproportion of oxidation state III be favored in Ru^{II}-OH_2/Ru^{IV}=O systems? *Journal of the American Chemical Society,* 2006, 128 (16), 5306-5307.

[56] Yagi, M., Syouji, A., Yamada, S., Komi, M., Yamazaki, H. and Tajima, S.: Molecular catalysts for water oxidation toward artificial photosynthesis. *Photochemical & Photobiological Sciences,* 2009, 8 (2), 139-147.

[57] Yamazaki, H., Shouji, A., Kajita, M. and Yagi, M.: Electrocatalytic and photocatalytic water oxidation to dioxygen based on metal complexes. *Coordination Chemistry Reviews,* 2010, 254 (21-22), 2483-2491.

[58] Chowdhury, A. D., Das, A., K, I., Mobin, S. M. and Lahiri, G. K.: Isomeric complexes of [Ru^{II}(trpy)(L)Cl] (trpy = 2,2′:6′,2″-terpyridine and HL = Quinaldic Acid): preference of isomeric structural form in catalytic chemoselective epoxidation process. *Inorganic Chemistry,* 2011, 50 (5), 1775-1785.

[59] Duan, L., Bozoglian, F., Mandal, S., Stewart, B., Privalov, T., Llobet, A. and Sun, L.: A molecular ruthenium catalyst with water-oxidation activity comparable to that of photosystem II. *Nature Chemistry,* 2012, 4 (5), 418-423.

[60] Hirahara, M., Nagai, S., Takahashi, K., Saito, K., Yui, T. and Yagi, M.: New series of dinuclear ruthenium(II) complexes synthesized using photoisomerization for efficient water oxidation catalysis. *Inorganic Chemistry,* 2015, 54 (15), 7627-7635.

In: Photoisomerization
Editor: Diego Sampedro
ISBN: 978-1-53615-313-2

Chapter 3

PHOTOISOMERIZATION AND PROPERTIES OF PHOTOISOMERS OF CARBOCYANINE DYES

Pavel G. Pronkin and Alexander S. Tatikolov*
N. M. Emanuel Institute of Biochemical Physics,
Russian Academy of Sciences, Moscow, Russia

ABSTRACT

One of the important features of polymethine (cyanine) dyes is isomerization about one of C–C bonds of the polymethine chain. Upon photoexcitation, polymethine dyes are capable of forming photoisomers, which are unstable and undergo back isomerization to form initial isomers. Usually initial (stable) isomers of polymethine dyes are *trans* (*EEEE*) isomers, whereas photoisomers (unstable) are in the mono-*cis* form. The dyes with a *meso*-substituent in the polymethine chain are characterized by dynamic *cis–trans* equilibrium, which depends on the polarity of the medium. Furthermore, a *meso*-substituent has steric effects on the configuration and energy of the dye molecule isomers, which is reflected in the kinetic characteristics of isomerization processes. In this chapter, spectral properties of the isomers, photoisomerization and thermal back isomerization of carbocyanine dyes, mostly *meso*-substituted

* Corresponding Author's Email: tatikolov@mail.ru.

carbocyanine dyes, are considered. *meso*-Alkyl-substituted thiacarbocyanine dyes are present in polar solvents as *cis* isomers and, hence, exhibit no photoisomerization, whereas in nonpolar solvents, in which the dyes are in the *trans* form, photoisomerization takes place. In contrast, the *meso*-substituted dyes 3,3'-dimethyl-9-phenylthiacarbocyanine and 3,3'-diethyl-9-(2-hydroxy-4-methoxyphenyl)thiacarbocyanine occur as *trans* isomers and exhibit photoisomerization in both polar and nonpolar solvents. The behavior of these dyes may be explained by the fact that the phenyl ring of the substituent in their molecules can be twisted at some angle, removing the substituent from the plane of the molecule and reducing its steric effect on the conformation of the *trans* isomer. In some cases, photoisomerization of *cis* isomers of *meso*-substituted carbocyanine dyes is also observed (for some *meso*-alkyl-substituted dyes complexed with DNA and chondroitin-4-sulfate; for 3,3'-diethyl-9-methoxythiacarbocyanine in solvents of moderate polarity). It is concluded that the cycle photoisomerization–thermal back isomerization of polymethine dyes can be used in various systems of information storage and deserves further investigation using modern research methods.

Keywords: polymethine (cyanine) dyes, photoisomerization, thermal back isomerization, *trans* and *cis* isomers, *meso*-substituted cyanines

Introduction

At present, polymethine (cyanine) dyes attract great attention of researches in different fields of science, technology engineering, pharmacology and medicine. This is due to the unique properties of polymethine dyes, which are determined by the key features of their molecular structure.

Polymethine dyes have been known for more than 150 years. The first polymethine dye was synthesized in 1856 and became one of the first synthetic dyes. Almost half a century polymethine dyes were of only theoretical interest. Due to the lack of photostability, these dyes did not find practical application for a long time. At the beginning of 20th century polymethine dyes found wide application in photography as spectral sensitizers to expand the photosensitivity spectrum of silver halide emulsions to the red and far-red regions [1].

The progress in science and technology stimulated application of polymethine dyes in various fields. In particular, these dyes are of interest for studying the correlations between the structure and the properties of molecules [1–3] and still attract the attention of researchers. At present, polymethine dyes are widely used in laser technology (for creating tunable laser media in the near IR spectral range, passive Q-switches of lasers and other devices of quantum electronics and optoelectronics) [3–5]. Polymethine dyes also attract interest due to the prospects of their use as spectral-fluorescent probes and labels in the study of various biological systems [6, 7], and in biomedicine for the purposes of visualization of tissues and photodynamic therapy [8, 9]. Various fields of applications of cyanine dyes determine the need for studying their photophysical and photochemical properties in solutions and molecularly organized media. This review chapter will be informative for researchers in photophysics and photochemistry of cyanine dyes and might be also interesting and useful for specialists in chemistry, physics and many interdisciplinary fields.

Polymethine (cyanine) dyes contain conjugated polymethine chains (with an odd number of CH groups) with terminal heteroatoms or heterocyclic groups. The dye molecule usually carries a positive charge distributed over the polymethine chain. Due to complete π-electronic conjugation in the polymethine chain, polymethine dyes have high extinction coefficients in the spectral range from 340 to 1400 nm [1–3]. The lack of rigidity in the molecules of these compounds determines the possibility of isomerization due to rotation around the C–C bonds of the polymethine chain. In this case, for the dyes, which do not have substituents in the *meso* position of the polymethine chain, as a rule, the all-*trans* (*EEEE*) isomer is thermodynamically more stable [10–17], and these dyes occur in solution just in the form of all-*trans* isomer. The introduction of substituents (alkyl groups, heteroatoms) in the *meso* position of the polymethine chain creates steric hindrances to the formation of the *trans* isomer, and in solutions of the substituted dyes a simultaneous presence in comparable concentrations of the all-*trans* and mono-*cis* isomers is observed [2, 18, 19].

Due to the lack of rigidity of the molecule, which is characteristic of cyanine dyes with an open polymethine chain, various vibrations and rotations of individual fragments are possible in their molecules, first of all, torsions and rotations around various bonds of the polymethine chain of cyanines. This determines short lifetimes (from tens to hundreds of picoseconds) of their excited singlet state (S_1), as well as low values of the quantum yields of fluorescence and intersystem crossing to the triplet state ($S_1 \rightarrow T$) [1, 2]. For polymethine dyes, as a rule, *trans*→*cis* (or *cis*→*trans*) photoisomerization is a potent nonradiative channel for deactivation of the excited state of the molecules. Hence, the study of photoisomerization of polymethine dyes is a key to understanding the photophysical and photochemical processes occurring in their molecules after photoexcitation. At present, there is a wide variety of polymethine dyes: cationic dyes—monomethine cyanines, carbocyanines (or trimethine cyanines), dicarbocyanines (pentamethine cyanines), tricarbocyanines (heptamethine cyanines), etc., as well as anionic (in particular, oxonols) and uncharged dyes (in particular, merocyanines and ketocyanines). In this chapter, we confine ourselves to the most common class of polymethine dyes—carbocyanines with a positively charged chromophore. Note that the total charge of such dyes can be not only positive, but also negative (or neutral) due to the presence of anionic substituents.

STRUCTURE AND SPECTRAL-FLUORESCENT PROPERTIES OF POLYMETHINE DYES

The general structure of polymethine (cyanine) dyes can be shown as follows:

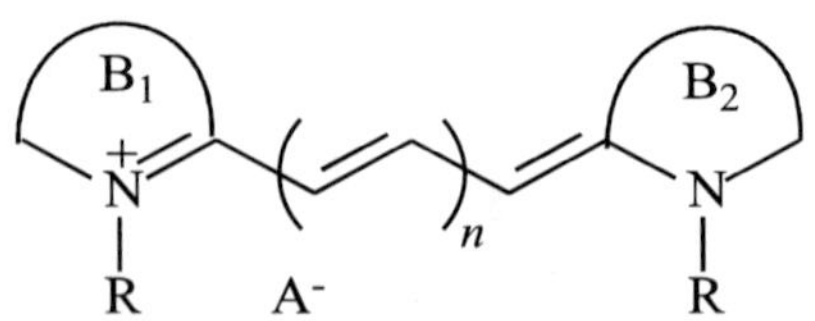

where n = 0, 1, 2, 3,…; nitrogen atoms can be a part of amino groups or terminal heterocycles B_1 and B_2; A^- = I^-, ClO_4^-, or anions of other acids.

For carbocyanines, $n = 1$. Quinolinyl, indolenyl, benzimidazolyl, benzothiazolyl, benzoxazolyl, and other heterocyclic residues can serve as B_1 and B_2 [1, 2].

Note that symmetric polymethine dyes are characterized by equalization of C–C bonds in the chromophore and approaching of their orders to one-and-a-half (polymethine state). Indeed, quantum mechanical calculations [20–23] have shown that the electronic structure of the polymethine chain in such dyes approaches the ideal polymethine state. From symmetry considerations, it follows that for symmetric cyanines ($B_1 = B_2$) this should lead to a uniform distribution of the electronic density along the polymethine chain and identical charges on the terminal groups. Hence, the electronic structure of symmetric cationic polymethine dyes can be represented by a superposition of two resonance structures [24]:

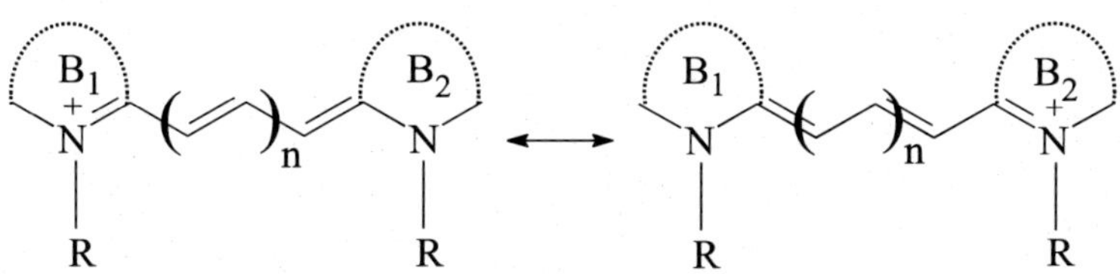

The orders of the bonds of the polymethine chain are the same (close to one-and-a-half) both in the ground state (S_0) and in the excited singlet state (S_1), whereas the signs of charges on the carbon atoms along the polymethine chain alternate: in the S_1 state, the charges on the carbon atoms of the polymethine chain change sign to the opposite [1–3], which determines the maximum change in the π-electron density for the electronic S_0–S_1 transition.

As noted above, cyanine dyes have intense narrow absorption bands (the extinction coefficient of the order of 10^5 L mol^{-1} cm^{-1}), the absorption maxima are located in the visible and near IR spectral regions. The spectral-luminescent properties of polymethine dyes are determined mainly by their long-wavelength π–π* electronic transition ($S_0 \leftrightarrow S_1$), whose dipole moment is directed along the long axis of the chromophore, i.e., along the polymethine chain of the dye from one terminal moiety of the molecule to another [2]. The ideal polymethine state is also characterized by the narrowest and most intense absorption bands, which determines the high selectivity of the spectral absorption and the brightness of the coloration of polymethine dyes [3]. However, in a number of cases these properties are related to the electronic transitions $S_0 \leftrightarrow S_n$, which are polarized at some angle to the $S_0 \leftrightarrow S_1$ transition and are usually localized in heterocyclic residues [2]. In the absorption spectra of most polymethine dyes, the intensity of the bands corresponding to the long-wavelength $S_0 \leftrightarrow S_1$ transition is substantially higher than for the short-wavelength $S_0 \leftrightarrow S_n$ transition. The presence of a system of alternating charges determines the high sensitivity of polymethine dyes to intermolecular interactions with the medium. This fact, as well as the weakened vibronic interactions, largely smooth out the vibronic structure in the electronic spectra of the dyes [2, 3]. Therefore, in their absorption spectra, as a rule, there is only one short-wavelength vibronic maximum as a characteristic shoulder at the edge of the main band.

Elongation of the polymethine chain in symmetric dyes by one vinylene group results in a bathochromic shift of the absorption maximum by about 100–130 nm [1, 2, 25]. Hence, by changing the number of units (n) in the polymethine chain, it is possible to obtain the dyes with absorption maxima in the range from 340 to 1400 nm [25, 26].

Isomerism and Photoisomerization of Polymethine Dyes

One of the main paths of degradation of the energy of the S_1 state of cyanines in low-viscosity media is usually *trans* → *cis* (or *cis* → *trans*) isomerization, performed by rotating a fragment of the dye molecule around one of the bonds of the polymethine chain by ~180° (also possible is internal conversion—the nonradiative transition $S_1 \rightarrow S_0$ induced by vibrations and torsions).

The existence of photoisomers of polymethine dyes was first proven in 1965 as a result of an experiment on flash photolysis of dye solutions [27]. The lack of rigidity leads to sufficiently high values of the photoisomerization quantum yields of cyanine dyes in polar solvents. For example, for the unsubstituted cationic dye 3,3'-dimethylthiacarbocyanine perchlorate (**1**), the photoisomerization quantum yield is 0.24 in dichloromethane [28]. Upon photoexcitation of its analogue, 3,3'-diethylthiacarbocyanine iodide (**2**) in polar solvents (ethanol, dimethyl sulfoxide, acetonitrile), the contributions of fluorescence and intersystem crossing to the triplet state upon deactivation of excited singlet molecules are low.

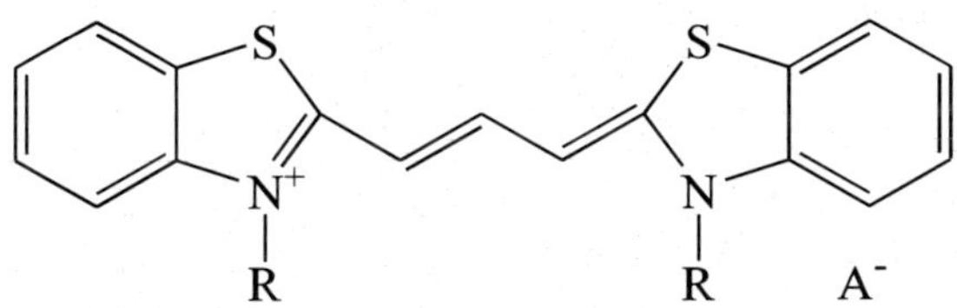

R = CH_3 A^- = ClO_4^- (**1**); R = C_2H_5 A^- = I^- (**2**)

The main process of nonradiative deactivation is the mechanism of *trans–cis* photoisomerization of dye molecules. Direct photoexcitation of polymethine dyes with no substituents in the polymethine chain (they are mainly present in solution in the form of *trans* isomers) leads to the formation of a short-lived *cis* photoisomer, which then undergoes back dark (thermal) isomerization to form the initial *trans* isomer (with a rate constant for **2** $k_i = 7 \times 10^2\ s^{-1}$ in ethanol [28]):

$$trans\text{-}S_0 + h\nu \rightarrow trans\text{-}S_1 \rightarrow cis\text{-}S_0 \xrightarrow{k_i} trans\text{-}S_0 \quad (1)$$

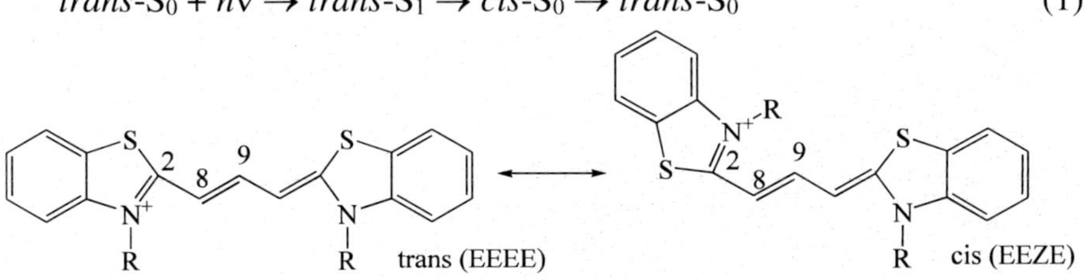

Upon flash photoexcitation of aerated solutions of the unsubstituted oxacarbocyanine dye 3,3'-diethyloxacarbocyanine iodide (**3**) in isopropanol and aqueous phosphate buffer solution (pH 7, 20 mmol L^{-1}), we observed the absorption and bleaching signals caused by photoisomerization and thermal back isomerization of the photoisomer formed [29].

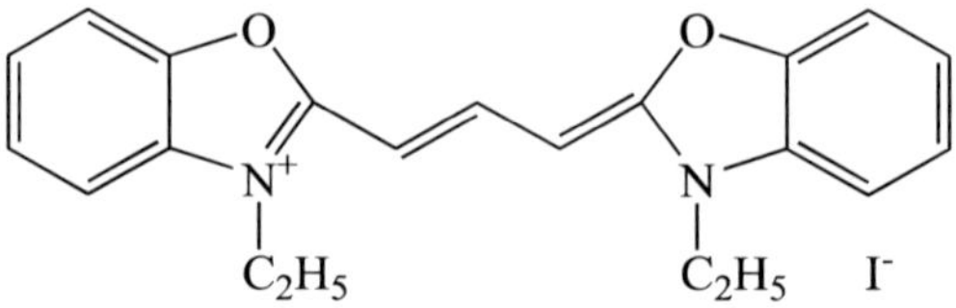

(3)

The criterion determining the assignment of the signals to the photoisomer was the absence of the effect of oxygen on the observed signals, since the photoisomers of cyanine dyes are formed, as a rule, from the S_1 state of the dyes having a too short lifetime (hundreds of picoseconds) to be quenched with atmospheric oxygen [30]. The bleaching maximum (λ_{max} = 480 nm) of the difference spectra of **3** approximately coincides with the absorption band of the dye (*trans* isomer), and the absorption maxima (λ_{max} = 505 nm) are in the long-wavelength region. The assignment of the spectrum to the *cis* isomer is based on [31]. The value of k_i for **3** does not depend on the excitation and registration wavelengths and is 9.0 s^{-1} (in aqueous phosphate buffer solution) [29].

As already noted, for the dyes that do not have substituents in the polymethine chain, as a rule, the all-*trans* (*EEEE*) isomer is thermodynamically more stable. These dyes are mainly present in solution in the form of this isomer [10]. Their photoisomers correspond to a *cis* conformation [32].

Potential Surfaces of Isomerization of Cyanine Dyes

Historically, several models of potential surfaces of cyanine isomerization (surfaces corresponding to isomerization rotation around one of the C–C bonds of the polymethine chain) were considered in the literature. One of the first models was the Rulliere model [33], in which the potential surfaces calculated by Orlandi and Siebrand for isomerization of stilbene

[34] were mechanically transferred to the case of isomerization of cyanines. This model was often used by other researchers (see, for example, [35, 36]). However, it was later shown [15] that the potential wave functions for polyenes and cyanines are of different nature, despite some similarity of the shapes of the potential surfaces for the S_1 state, so the results of Orlandi–Siebrand's calculations are not applicable for cyanines. Now in the literature, a model of potential surfaces for cyanine dyes is used, starting from which, irrespective of the solvent, only the surface of S_1 has a third minimum (turn by ~ 90°), which roughly corresponds to the top of the potential barrier of the surface of S_0 having two minima (Figure 1) [15]. Upon photoexcitation, a molecule in the form of a *trans* isomer passes to the central minimum of S_1, from which, when it reaches the surface of S_0, it slips down to a minimum of *cis*-S_0 (*trans–cis* photoisomerization) or *trans*-S_0 (nonradiative degradation of the excitation energy). If the excited molecule is in the form of a *cis* isomer, then it also passes to the central minimum of S_1, from which, when it reaches the surface of S_0, it slips down to a minimum of *cis*-S_0 (nonradiative degradation of the excitation energy) or *trans*-S_0 (*cis-trans* photoisomerization). The quantum yields of photoisomerization and nonradiative degradation depend on the position of the minimum of S_1 with respect to the maximum of S_0. Theoretical calculations by the CS INDO method [15] and experimental verification of the consistency of this model with the isomerization process [17] have shown its correspondence to cyanine isomerization.

For cyanines with no substituents in the polymethine chain, nonradiative relaxation of the excited singlet state (both *trans*- and *cis*-S_1), occurring by movement along the potential surface of isomerization, leads to both *cis* and *trans* isomers of the dye. Thus, for these dyes, photoisomerization occurs upon photoexcitation of both *trans* and *cis* isomers.

In the case of most *meso*-substituted thiacarbocyanines (first of all, alkyl-substituted), the minimum potential energy of the dye in the S_1 state corresponding to the perpendicular conformation is shifted with respect to the maximum of the energy barrier of the S_0 state toward the *cis* isomer (Figure 2) [15, 17].

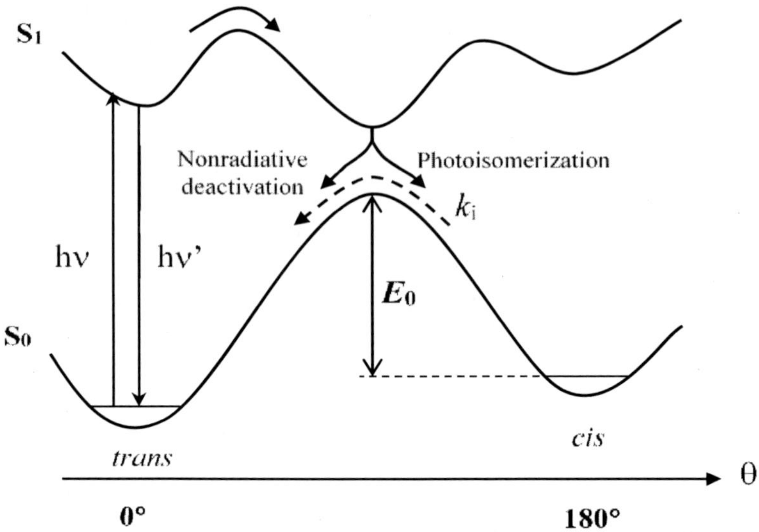

Figure 1. General scheme of the potential surfaces of isomerization of cyanine dyes (only S_0 and S_1 surfaces are shown; photoexcitation of the *trans* isomer is shown).

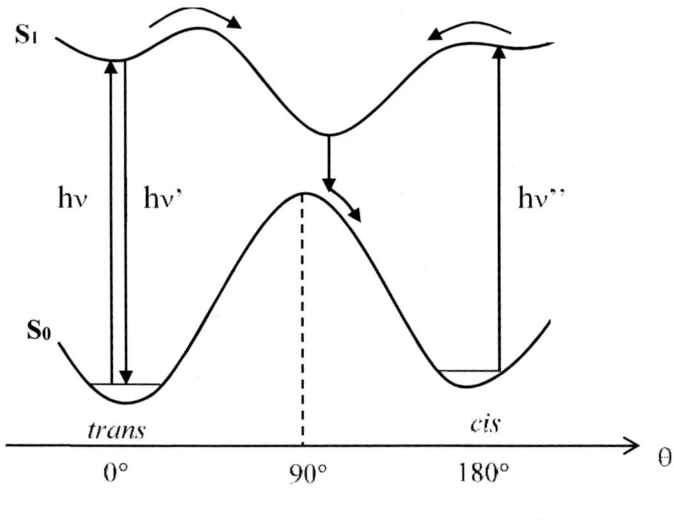

Figure 2. Scheme of the potential surfaces of isomerization of *meso*-substituted cyanine dyes (only S_0 and S_1 surfaces are shown).

For this reason, upon nonradiative relaxation of the S_1 state of such dyes by moving along the S_1 potential surface, the *trans* conformation of the dyes is not attained and photoexcitation into the absorption band of the *cis* isomer does not lead to the formation of the *trans* form, that is, the *cis* isomers of most *meso*-substituted thiacarbocyanines are not photoisomerized [28].

EFFECT OF SOLVENT VISCOSITY ON THE KINETICS OF PHOTOISOMERIZATION AND BACK (THERMAL) ISOMERIZATION

The kinetics of isomerization of cyanines in solution is determined, first of all, by a potential barrier of rotation around the isomerizing bond. For isomerization processes with relatively low activation energies close to the activation energy of the viscous flow of the solvent (this is usually isomerization from the S_1 state, that is, photoisomerization), it strongly depends on the viscosity of the medium (it slows down sharply with increasing viscosity) [32, 37–40]:

$$E_a = E_0 + aE_\eta, \tag{2}$$

where E_a is the apparent activation energy of isomerization, E_0 is the internal energy barrier of isomerization, E_η is the activation energy of a viscous flow, and a is a coefficient depending on the slope of the potential barrier in the transient state of the process ($0 \leq a \leq 1$). For photoisomerization, the coefficient a is much larger than for the thermal back isomerization (in the latter case, a weak dependence of E_a on the viscosity is observed, which is manifested only in highly viscous solvents).

Due to the very efficient photoisomerization process, the lifetime of the S_1 state of the dyes in low-viscosity solvents is much shorter than the radiative lifetime and is usually hundreds of picoseconds or less. In high viscosity media, internal rotations in the cyanine molecule are hindered, which leads to a drop in the quantum yield of photoisomerization and an increase in the quantum yield of fluorescence, the fluorescence lifetime, and the quantum yield of the triplet state, since fluorescence and intersystem crossing to the triplet state compete with photoisomerization.

In a number of works, using pico- and femtosecond spectroscopy, ultrafast processes of photoisomerization of a number of cyanines have been studied (see, for example, [42, 43]). The results indicate, in particular, that the nature of the reaction coordinate plays an important role in determining the dependence of photoisomerization on viscosity. A more detailed consideration of these works is beyond the scope of this chapter.

EFFECTS OF SOLVENT POLARITY ON THE ISOMERIZATION PROCESSES OF POLYMETHINE DYES

In polar solvents, polymethine dyes are present in the form of free ions (cations); in this case, the influence of polarity of the medium on the isomerization of polymethine dyes is small [17]. However, in weakly polar and nonpolar solvents, the dyes form ion pairs with a counterion, which can affect their photochemistry, in particular, the processes of photoisomerization and back isomerization of the resulting photoisomer.

In [44], photophysical and photochemical processes occurring upon flash photoexcitation in molecules of dyes **1** (iodide) and **2** (chloride) in a mixture of dimethyl sulfoxide (DMSO)–toluene of various compositions were studied. With a decrease in the DMSO content in the solution down to 3 vol %, the relative quantum yield of intersystem crossing of dye **1** to the triplet state increases by two orders of magnitude, which is accompanied by a decrease in the quantum yield of *trans–cis* photoisomerization. This effect is not observed for dye **2**. The authors explain these results by the heavy atom effect upon the formation of ion pairs of the dye cation **1** with the iodide counterion under nonpolar conditions.

In our work [45], we studied the effect of counterions (ClO_4^-, Cl^-, I^-) on photochemical processes (including isomerization) in molecules of cationic polymethine dyes (in particular, benzimidacyanine dye **4**) in acetonitrile–toluene mixtures.

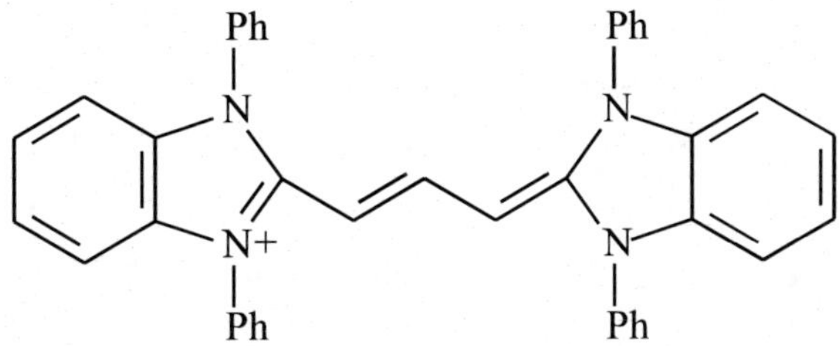

(4)

In low polarity media, in which the dyes formed ion pairs with counterions, an increase in the rate constant and a decrease in the activation energy of back isomerization of benzimidacyanines were observed. The effect was explained by the electrostatic influence of the counterion in ion pairs on the electronic density distribution in the dye cation, leading to a decrease in the activation barrier of the process. The influence of the counterion on the transient ("twisted") state of isomerization, which lowers its energy, is also possible.

Isomerism and Photoisomerization of *Meso*-Substituted Carbocyanine Dyes

While carbocyanine dyes, which do not have *meso* substituents in the polymethine chain, are present in solutions as *trans* isomers, *meso*-substituted thiacarbocyanine dyes in the crystalline state are usually present as *cis* isomers [18].

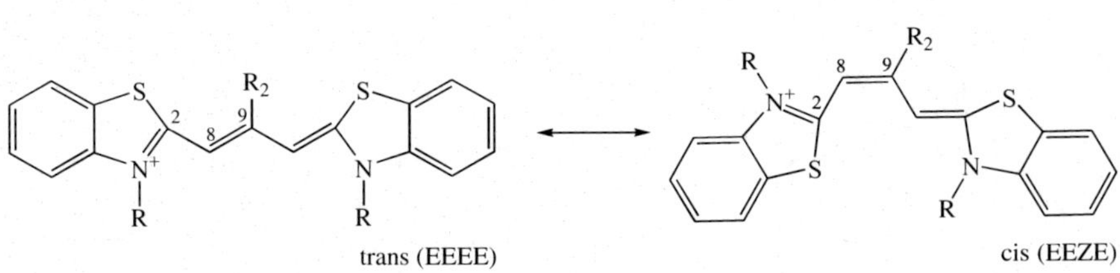

Bulky substituents in the *meso* position of the polymethine chain create significant steric hindrances to the formation of the *trans* isomer; in solutions of such compounds, simultaneous presence (in comparable concentrations) of all-*trans* and *cis* isomers is observed [18, 19]. In particular, we have shown that 3,3’,9-trimethylthiacarbocyanine iodide (**5**; Cyan 2), having a *meso*-methyl substituent in the polymethine chain, is mainly in the *cis* form in polar solvents [46].

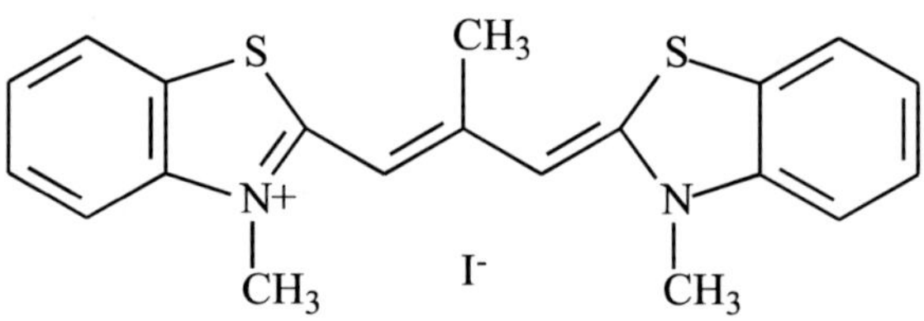

(**5**)

At low temperatures in solutions, the absorption bands of *meso*-substituted thiacarbocyanines are split into two closely spaced bands [19]. The equilibrium position also depends strongly on the solvent: *trans* isomers predominate in low polarity media, while in highly polar solvents they are present only as a small impurity to the *cis* isomer [10, 18, 19, 47, 48].

An increase in the effective volume of the substituent in the *meso* position of the polymethine chain of the molecule promotes an increase in the *cis* isomer content. It has been established that 9-tert-butyl-3,3'-diethylthiacarbocyanine cation (**6**) exists both in solution and in the solid phase exclusively as a nonplanar di-*cis* (*EZZE*) form [10, 47].

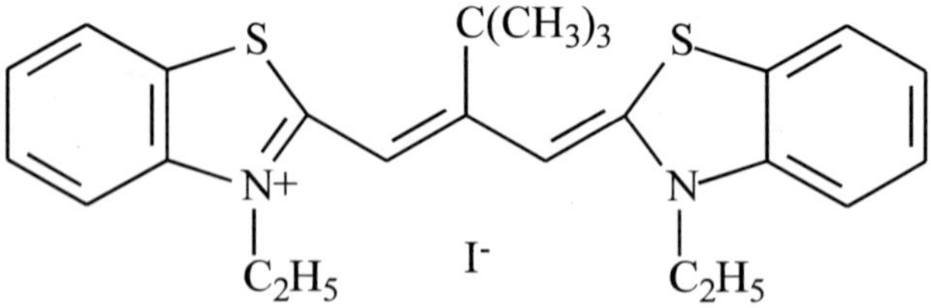

(**6**)

The simultaneous presence of *trans* and *cis* isomers in dye solutions is explained by distortion of the planar structure of the dye made by substituents in the *meso* position of the polymethine chain of the dyes. This leads to a significant increase in the energy of the *trans* isomer, which becomes close to that of the *cis*-isomer [14, 18–20]. The dynamic equilibrium observed depending on the solvent polarity, is due to the electrostatic interaction of the dye cation with a counterion in low polarity media. The formation of ion pairs in low polarity media leads to a decrease in the energy of the *trans* isomer and its preferential formation, whereas in polar media the *cis* isomer is formed, which has a lower energy [20]. The different isomers present in the solution cause a difference in the photochemical properties of *meso*-substituted carbocyanine dyes in solvents of different polarity. In polar solvents, where the dyes are predominantly in the form of *cis* isomers, photoisomerization is generally not observed, while in low polarity solvents, in which the dyes are present as *trans* isomers, photoisomerization to form *cis* isomers occurs upon photoexcitation.

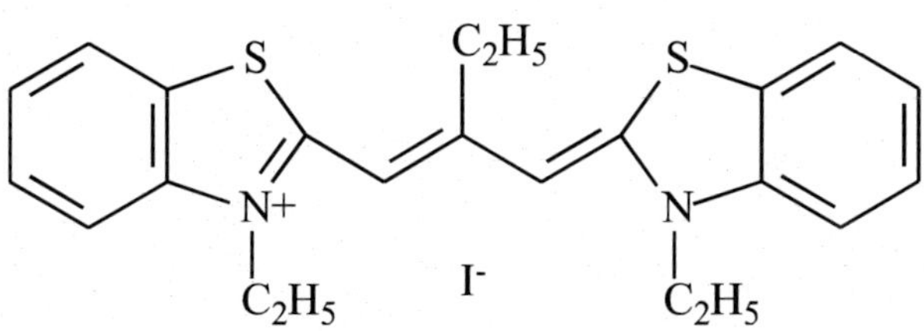

(**7**)

This rule is fulfilled by a series of thiacarbocyanines with *meso*-alkyl substituents. In particular, we have shown that 3,3',9-triethyl-thiacarbocyanine iodide (**7**) in isopropanol is in the form of a *cis* isomer, and in dioxane in the form of a *trans* isomer [48].

Upon flash photoexcitation, the signals corresponding to photoisomerization processes were not detected in the former case (i.e., the *cis* isomer was not photoisomerized), whereas in the latter case, the signals corresponding to the formation and subsequent decay of the *cis* photoisomer were detected (that is, photoisomerization of the *trans* isomer was observed;

see insets in Figure 3) [49]. By adding the resulting difference absorption spectrum of the photoisomer (Figure 3, curve *2*) to the absorption spectrum of the initial (*trans*) form of the dye (Figure 3, curve *1*), we obtained a stationary absorption spectrum of the photoisomer (Figure 3, curve *3*).

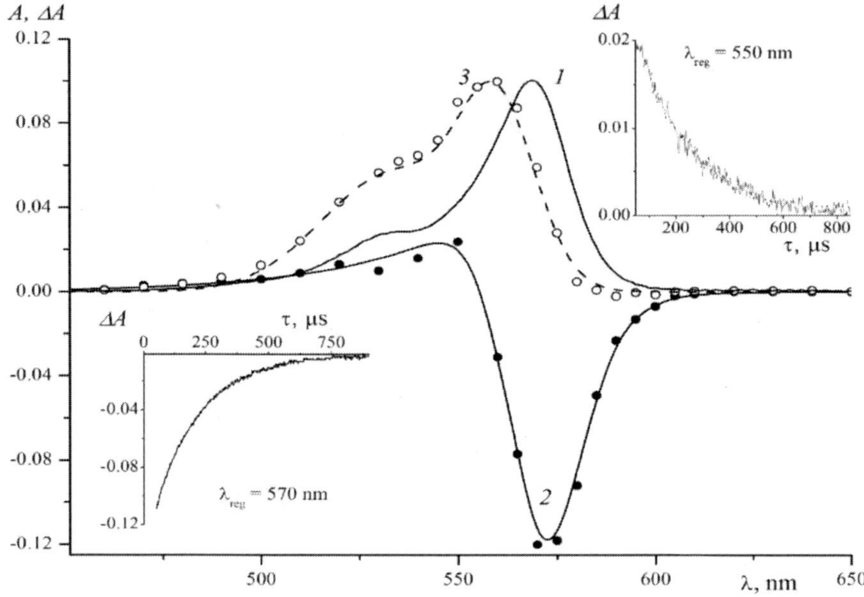

Figure 3. Absorption spectrum of dye **7** in dioxane (*1*; $c_{dye} = 9.7 \times 10^{-7}$ mol L^{-1}), difference absorption spectrum of the dye photoisomer in dioxane obtained by flash photolysis (*2*), and reconstructed stationary absorption spectrum of dye **7** photoisomer (*3*). The insets show the kinetics of the signal due to back isomerization of **7** (λreg = 550 nm and 570 nm).

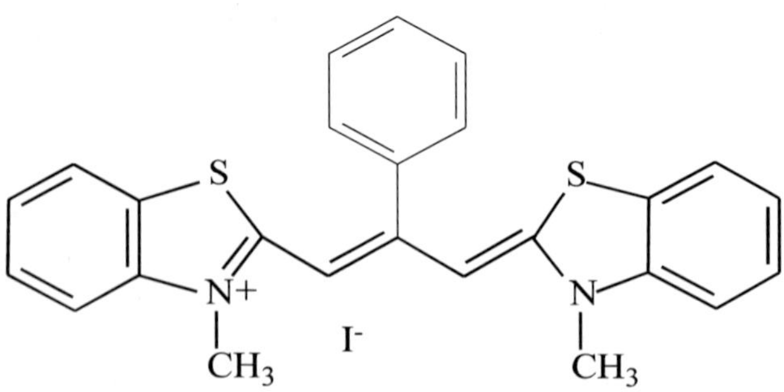

(**8**)

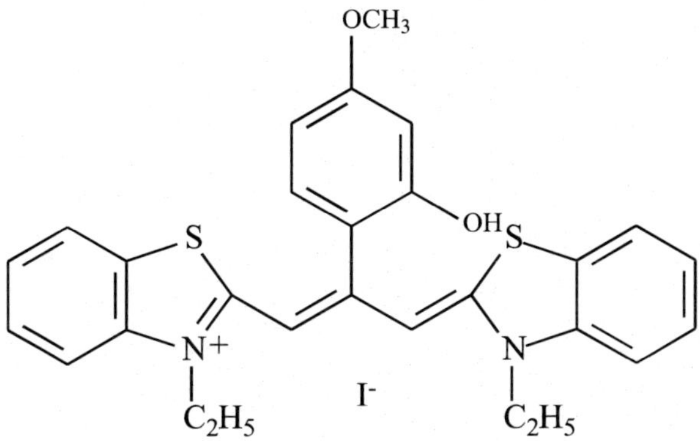

(**9**)

However, for dyes with other *meso* substituents this rule is often not fulfilled. In the case of two *meso*-substituted thiacarbocyanine dyes, 3,3'-dimethyl-9-phenylthiacarbocyanine iodide (**8**) and its analogue, 3,3'-diethyl-9-(2-hydroxy-4-methoxyphenyl)thiacarbocyanine iodide (**9**), the analysis of the absorption, fluorescence, and fluorescence excitation spectra has shown that these dyes in solutions, regardless of the polarity of the solvent, are in the form of *trans* isomers (Figure 4) [50].

Apparently, for dyes **8** and **9**, the substituents in the *meso* position of the polymethine chain introduce less steric distortion into the structures of the *trans* isomers than alkyl substituents in the *meso*-alkyl-substituted dyes.

This is probably explained by twisting of the phenyl *meso* substituents in **8** and **9** relative to the dye chromophore plane, which sharply reduces their distorting effect on the planar structure of the dye chromophore and, hence, on the relative energy of the isomers. As a result, photoisomerization is observed for these dyes even in polar solvents. In particular, in the case of flash photoexcitation of **8** and **9** in ethanol, the difference spectra of *cis* photoisomers are observed (Figure 5, curve *2*), the bleaching maxima of which (λ = 560 and 568 nm, respectively) approximately coincide with the positions of the maxima in the stationary absorption spectra of the dyes in ethanol (Figure 5, curve *1*), and the absorption maxima are observed in both the long-wavelength (λ = 605–610 nm), and short-wavelength regions (λ = 490–510 nm) [50].

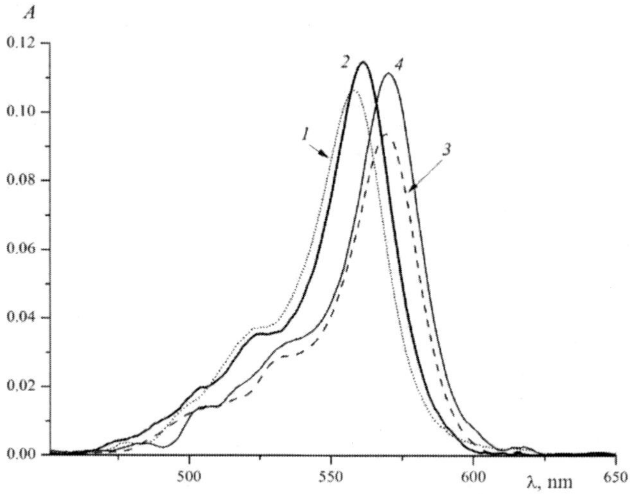

Figure 4. Absorption spectra of dye **8** ($c_{dye} = 5.85 \times 10^{-7}$ mol L^{-1}) in acetonitrile (*1*), ethanol (*2*), DMSO (*3*), and dioxane (*4*).

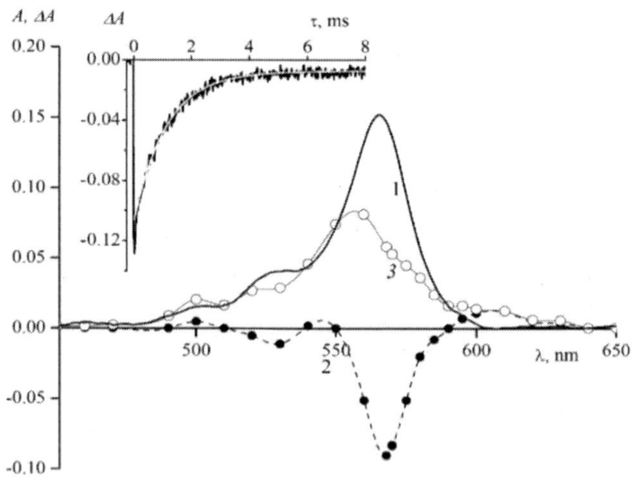

Figure 5. Difference absorption spectrum of the photoisomer of dye **8** in ethanol (*2*; $c_{dye} = 1.23 \times 10^{-6}$ mol L^{-1}) measured by flash photolysis technique, absorption spectrum of the original dye (*1*), and reconstructed stationary absorption spectrum of the *cis* photoisomer (*3*). The inset shows the kinetics of the signal due to back isomerization of **8** ($\lambda_{reg} = 560$ nm).

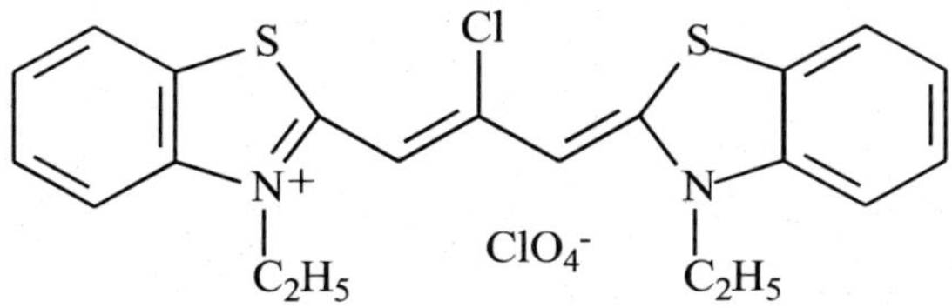

(**10**)

Reconstruction of the spectra of the photoisomers using stationary absorption spectra of the dyes (*trans* isomers) (Figure 5, curve *1*) permitted obtaining stationary absorption spectra of the *cis* photoisomers of **8** and **9** (Figure 5, curve *3*). As in the case of *meso*-unsubstituted dye **2**, which forms *cis* photoisomer upon photoexcitation [28, 31], the stationary absorption spectra of the photoisomers of dyes **8** and **9** are shifted to the short-wavelength side (λ_{max} ~ 550 and 558 nm, respectively).

We showed that dye **10**, which has Cl as a substituent in the *meso* position of the polymethine chain, is in the form of a *cis* isomer in water, and in the form of a *trans* isomer in a less polar solvent, isopropanol. Upon flash photolysis of **10** in isopropanol, signals corresponding to the formation of a *cis* photoisomer were obtained [49]. The decay rate constant for the photoisomer of **10** was 1.6×10^3 s^{-1}. Upon flash photoexcitation of aqueous solutions of *meso*-substituted oxacarbocyanine dyes 3,3'-dimethyl-9-ethyloxacarbocyanine (**11**) and 3,3',9-triethyl-5,5'-dimethyloxacarbocyanine (**12**), the formation of *cis* photoisomers was also detected [51]. The intensity of the signals of photoisomers of **11** and **12** is much lower than for their unsubstituted analogue **3**, being in the *trans* form, which can be explained by the presence of a significant admixture of *cis* isomers of the dyes (in equilibrium with *trans* isomers), whose photoisomerization does not occur [18, 19].

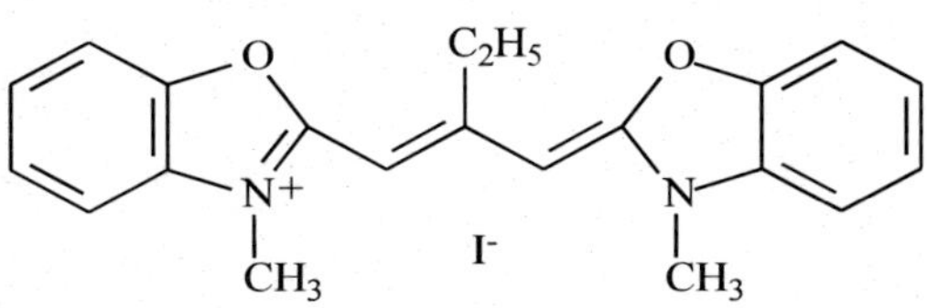

(**11**)

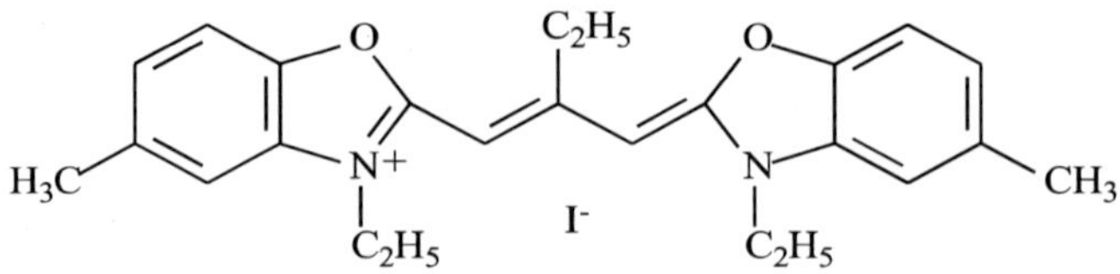

(**12**)

However, for some *meso*-substituted thiacarbocyanines, photoisomerization of both *trans* and *cis* isomers is observed. We have shown that 3,3'-diethyl-9-methoxythiacarbocyanine **13** in solvents of moderate polarity (ethyl acetate, isopropanol) is in the form of an equilibrium mixture of *cis* and *trans* isomers.

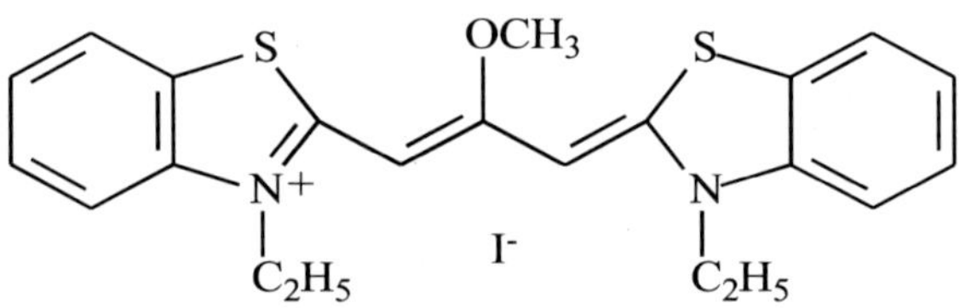

(**13**)

Depending on the photoexcitation wavelength, photoisomerization of either *cis* or *trans* isomer was observed, leading to the formation of *trans* or *cis* photoisomer, respectively [49]. With selective excitation by a dye-tunable nitrogen laser (λ_{ex} = 540, 520, 510, and 470 nm) into the absorption bands of *cis* or *trans* isomers of dye **13** in ethyl acetate and isopropanol, the corresponding difference absorption and bleaching spectra of *trans* or *cis* photoisomers were obtained.

In particular, upon photoexcitation of **13** in ethyl acetate at wavelengths of 540 and 510 nm, the maximum bleaching (λ_{max} = 535 nm) approximately coincides with the long-wavelength peak in the stationary absorption spectrum of dye **13** (the *trans* isomer is consumed), and the absorption maximum (λ_{abs} = 490 nm) is close to the absorption spectrum of the *cis* form (λ_{max} = 498 nm; the *cis* isomer is formed). Upon excitation of **13** with light at a wavelength of 520 or 470 nm, the maximum bleaching (λ_{max} = 500 nm) coincides with the maximum of the *cis* isomer in the absorption spectrum of the dye (the *cis* isomer is consumed), and the absorption maximum (λ_{abs} = 525 nm) with the maximum of the *trans* form (the *trans* isomer is formed). [49]. Transient signals due to the *trans–cis* (or *cis–trans*) isomerization processes were observed, with the observed decay rate constant of 1.35×10^6 and 1.7×10^6 s^{-1} in ethyl acetate and isopropanol, respectively. Since the *cis* and *trans* isomers in the solution of dye **13** are in equilibrium, the observed rate constant of the signal decay corresponds to the sum of the rate constants of *cis* → *trans* and *trans* → *cis* isomerization.

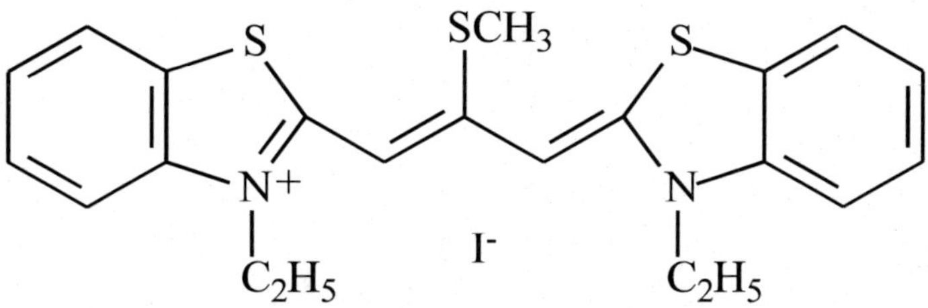

(**14**)

The rate constants of thermal *cis–trans* transitions for *meso*-substituted thiacarbocyanine dyes were found to be much higher than for their unsubstituted analogue **2** (250 s^{-1} in ethanol) [28]. In particular, the kinetics of the transition of the *cis* photoisomers of **7** and **14** to the initial *trans* forms obey the first order law with rate constants $k_i \sim 5 \times 10^3$ and 5×10^5 s^{-1}, respectively [49].

Such an increase in the rate constants of the thermal back isomerization is due to steric hindrances, which create bulky groups of substituents in the *meso* position of the polymethine chain. The steric influence of the *meso* substituents leads to removing terminal heterocyclic groups of dye molecules from the plane, which results in distortion of the valence angles, an increase in the energy of the *cis* and *trans* isomers, and a corresponding decrease in the potential barrier of thermal isomerization in the ground state [28]. Similar effects were observed for *meso*-substituted oxacarbocyanine dyes **11** and **12**: the values of k_i for **11** and **12** were three orders of magnitude higher (1.5×10^4 and 1×10^4 s^{-1}) than for their unsubstituted analogue (9.0 s^{-1}) [51].

For thiacarbocyanine dyes **8** and **9** with bulky *meso* substituents based on phenyl rings, the values of the observed rate constants for the decay of *cis* photoisomers were found to be 340 and 865 s^{-1} for **8** and **9**, respectively (in ethanol) [50]. These constants were rather close to k_i for the unsubstituted dye **2** (250 s^{-1}), indicating a weak steric effect of *meso* substituents in **8** and **9** on the configuration of the *cis* photoisomer. This can be explained by the possibility of turning the phenyl ring of the substituent, which removes the substituent from the chromophore's plane of the molecule and reduces its steric effect on the configuration of the isomers.

Photoisomerization and Back Isomerization of Photoisomers of Polymethine Dyes in Structurally Organized Media

It is known that polymethine dyes can form noncovalent complexes with surfactants, polyelectrolytes, biomacromolecules. The formation of such complexes, as a rule, hinders the intramolecular degrees of freedom of dye molecules leading to nonradiative dissipation of excitation energy, which results in a drop in the quantum yield of photoisomerization and an increase in the quantum yield of fluorescence [6, 7]. In the case of *meso*-substituted polymethine dyes, which are characterized by a dynamic *cis–trans* equilibrium depending on the medium, the formation of such complexes can significantly affect the *cis–trans* equilibrium.

In [52], photophysical and photochemical processes occurring in molecules of *meso*-substituted thiacarbocyanine dyes 3,3'-bis(γ-sulfopropyl)-5-methoxy-4',5'-benzo-9-ethylthiacarbocyanine (**15**), 3,3'-bis(γ-sulfopropyl)-6,6'-dimethyl-9-ethylthiacarbocyanine (**16**), and 3,3'-bis(γ-sulfopropyl)-5-methyl-5'-phenyl-9-ethylthiacarbocyanine (**17**) in aqueous solutions in the presence of cationic (cetyltrimethylammonium bromide), anionic (sodium dodecyl sulfate, SDS) and neutral (Triton X-100) surfactants were studied.

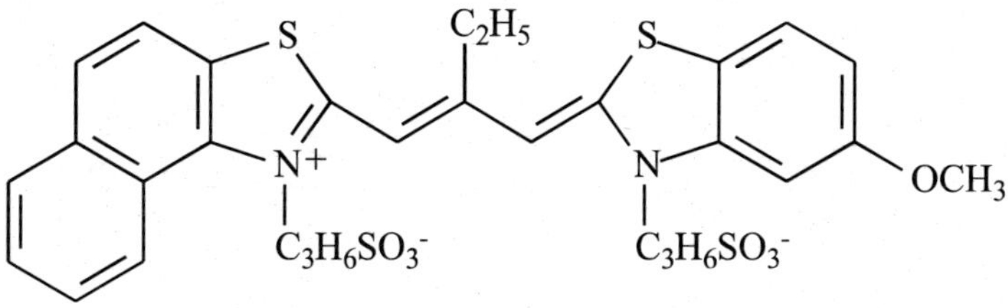

(**15**)

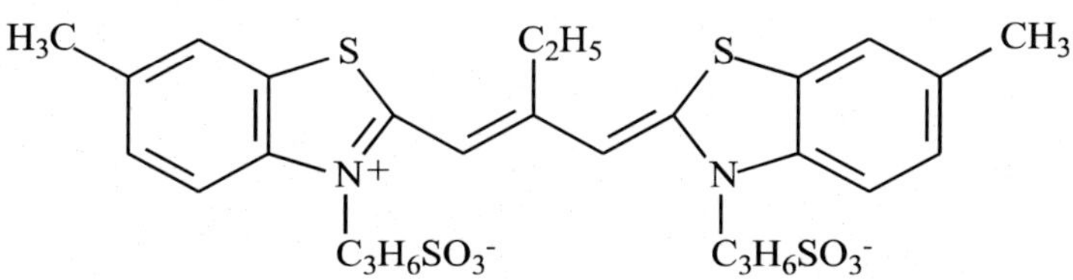

(**16**)

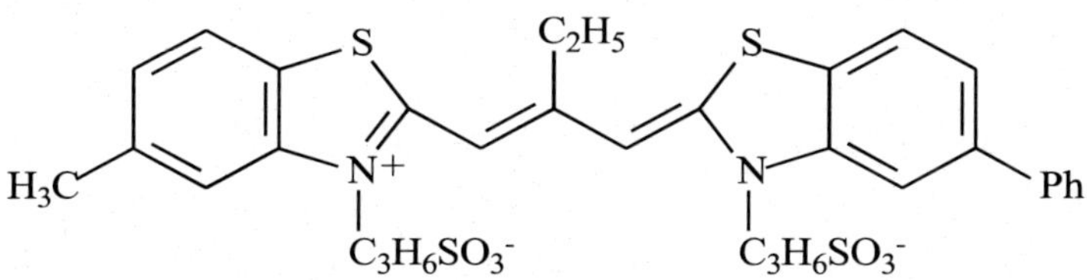

(**17**)

In an aqueous solution in the absence of surfactants, dyes **15–17** are in the form of dimers in equilibrium with *cis* monomers and are not capable of photoisomerization. At surfactant concentrations above critical micelle concentration (CMC), the dimers decompose into monomers, and *cis* monomers are converted into the *trans* form upon solubilization of the dye molecules in the surfactant micelles. The resulting *trans* isomers of the dyes are capable of photoisomerization with the formation of *cis* photoisomers.

We studied the behavior of *meso*-substituted polymethine dyes 3,3'-di-(γ-sulfopropyl)-4,5,4',5'-dibenzo-9-ethylthiacarbocyanine betaine (**18**,

DEC) and **5** in micellar systems of biological surfactants—bile salts sodium cholate, sodium deoxycholate, and sodium taurocholate [53].

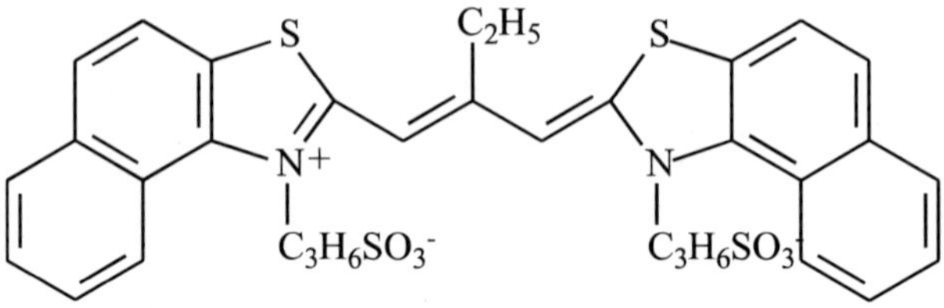

(**18**)

Upon binding to micelles of bile salts, as in the case of synthetic surfactants, these dyes are converted into a *trans* form. Upon flash photolysis of aerated aqueous solutions of the dyes in the presence of micelles of bile salts and SDS, signals caused by photoisomerization of the *trans* isomers of the dyes and thermal back isomerization of the *cis* photoisomers formed were observed. The lifetimes of photoisomers of **5** and **18** in the presence of bile salt micelles were in the range of 60–190 μs. In deoxygenated solutions, the appearance of the triplet state of the dyes was also observed.

In [54], photophysical and photochemical properties of thiacarbocyanine dyes **2**, **19**, and **20** were studied in the presence of water-soluble polyelectrolytes: polystyrene sulfonate (PSS), polyacrylic acid, (PAA), and polymethacrylic acid (PMA). For **2** in the presence of PSS, a significant (tenfold) decrease in the quantum yield of *trans–cis* photoisomerization (accompanied by an increase in fluorescence and a moderate increase in the intersystem crossing to the triplet state) was observed.

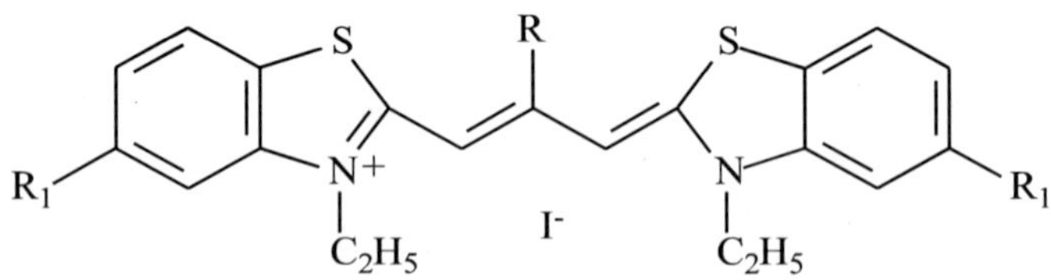

R= CH_3, R_1 = H (**19**); R = C_2H_5, R_1 = Cl (**20**).

Besides surfactants and synthetic polyanions, cationic cyanine dyes are capable of forming noncovalent complexes with DNA. The interaction with DNA hampers the formation of photoisomers of *meso*-unsubstituted polymethine dyes. In [55], we reported a drop in the yield of the photoisomer of dye **2** upon the formation of its noncovalent complex with ds-DNA in a buffer solution. Upon flash photolysis of aerated solutions of *meso*-substituted thiacarbocyanine dyes **7–10** and **14** in the presence of DNA (with DNA concentration in the range of (2.3–5.0) × 10^{-4} mol L^{-1}), signals of photoisomers were not observed [49, 50, 56]. Photoisomerization was also hampered by noncovalent interaction of oxacarbocyanine dyes **3**, **11**, **12** with DNA [51]. Upon flash photolysis of aerated solutions of **11** and **12** with DNA at $c_{DNA} = 4.4 \times 10^{-5}$ mol L^{-1} and higher, the signals of photoisomers were not observed. For dye **3** with $c_{DNA} = 2.39 \times 10^{-4}$ mol L^{-1}, the absorption intensity of the photoisomer decreases 2.5 times, and its lifetime increases 5 times.

Thus, the interaction with DNA and synthetic polyanions hampers the formation of photoisomers in the case of a large number of cationic polymethine dyes, which is explained by the steric factor of complexation, which slows down the nonradiative processes of dissipation of excitation energy. Nevertheless, for some *meso*-substituted thiacarbocyanine dyes, the formation of photoisomers was enhanced upon the complexation. In particular, in [57], the photophysical and photochemical properties of the *meso*-methyl-substituted dye **19** were studied in aqueous solutions in the presence of PSS, PAA and PMA. Flash photolysis of **19** in the presence of PAA and PMA revealed a weak signal due to *cis*→*trans* photoisomerization of the dye bound to the polyelectrolyte in the *cis* form and subsequent dark decay of the resulting *trans* photoisomer (τ = 700 μs at 560 nm). This contradicts the rule that *cis* isomers of *meso*-alkyl-substituted carbocyanines are not photoisomerized (dye **19** is not photoisomerized in an aqueous solution without polyelectrolytes). Similarly, in our work [46], we showed the enhancement of *cis*→*trans* photoisomerization of *meso*-methyl-substituted dyes **5** and **19** by noncovalent binding to DNA and chondroitin-4-sulfate in a buffer solution. Upon flash photolysis of **5** and **19** in the presence of DNA and chondroitin-4-sulfate, signals due to the formation of *trans* photoisomers and their subsequent dark decay (τ = 3.9 and 2.2 ms for **5** in the presence of DNA and chondroitin-4-sulfate, respectively; τ = 0.7 ms for **19** in the presence of biopolymers, λ_{reg} = 560 nm) were observed. To explain the effect of stimulation of photoisomerization of *meso*-substituted *cis* isomers by complexation with polyelectrolytes, an assumption was made that the biopolymer matrix of polyelectrolyte affects the potential surfaces of photoisomerization of the dyes: deformation of the potential surface of S_1 occurs with a shift of its minimum toward the *trans* isomer [46, 57].

CONCLUSION

The use of polymethine dyes in various fields of science and technology requires a detailed study of their photophysical and photochemical properties. Recently, a lot of works have been published on photophysics and photochemistry of polymethine dyes (see, for example, [58–61]). Photoisomerization is one of the main properties of polymethine dyes; at present, it is being studied in detail with the use of modern micro-, nano-, pico- and femtosecond techniques. However, until recently, this property of polymethine dyes has not been used in practice. On the contrary, rigidization of the polymethine chain of dyes is often carried out by introducing “bridges” that suppress internal conversion and photoisomerization processes to increase the stability of the dyes and the fluorescence quantum yield [62]. It should be noted here that the cycle photoisomerization–reverse dark isomerization of the photoisomer is completely reversible and, ideally, does not lead to decomposition of the dye molecule. Since the lifetime of the photoisomer, depending on the structure of the dye and the medium, can vary in very wide range (from seconds to picoseconds), this cycle can be used for transient recording and storage of information within different time ranges in modern information recording systems and devices. So we may expect in future widespread use of polymethine dyes as information storage media in various fields of science and technology.

Of great importance is also the dynamic *cis–trans* isomeric equilibrium, which strongly depends on the molecular microenvironment, observed for *meso*-substituted polymethine dyes. Since the photophysical and photochemical properties of the *cis* and *trans* isomers of such dyes are very different, these dyes can be widely used as probes for studying various molecularly organized and nanostructured systems. Indeed, a number of *meso*-substituted polymethine dyes have already been used as spectral-fluorescent probes for biomacromolecules (DNA, albumins, etc.) [63–67]. Next in turn is the use of polymethine dyes as kinetic probes, as it has been found recently that the lifetime of the photoisomers of *meso*-substituted oxonols (anionic polymethine dyes) [68] and squarylium cyanines [69] depend strongly on the polarity of the molecular microenvironment.

In summary, it can be noted that the study of photoisomerization and the properties of photoisomers of polymethine dyes using modern research methods is still relevant both for theory and for the practical application of polymethine dyes.

This work was supported by the Russian Foundation for Basic Research (project no. 16-03-00735).

References

[1] Mishra, A.; Behera, R. K.; Behera, P. K.; Mishra, B. K.; Behera, G. B. *Chem. Rev.* 2000, vol. 100, 1973–2011.

[2] Ishchenko, A. A. *Stroenie i spektral'no-luminestsentnye svoistva polimetinovykh krasitelei* [*Structure and Spectral-Luminescent Properties of Polymethine Dyes*]; Naukova Dumka: Kiev, 1994; 231 p.

[3] Davidenko, N. A.; Ishchenko, A. A.; Kuvshinsky, N. G. *Fotonika molekulyarnykh poluprovodnikovykh kompozitov na osnove organicheskikh krasitelei* [*Photonics of Molecular Semiconductor Composites Based on Organic Dyes*]; Naukova Dumka: Kiev, 2005; 296 p.

[4] Pierce, B. M. *Proc. SPIE* 1991, vol. 1560, 148–161.

[5] Pittman, M.; Plaza, P.; Martin, M. M.; Meyer, Y. H. *Opt. Commun.* 1998, vol. 158, 201–212.

[6] Levitus, M.; Ranjit, S. *Q. Rev. Biophys.* 2011, vol. 44, 123–151.

[7] Tatikolov, A. S. *J. Photochem. Photobiol. C: Photochem. Reviews* 2012, vol. 13, 55–90.

[8] Pandey, R. K.; James, N. S.; Chen, Y.; Dobhal, M. P. *Top. Heterocycl. Chem.* 2008, vol. 14, 41–74.

[9] James, N. S.; Chen, Y.; Joshi, P.; Ohulchanskyy, T. Y.; Ethirajan, M.; Henary, M.; Strekowski, L.; Pandey, R. K. *Theranostics* 2013, vol. 3, 692–702.

[10] Henrichs, P. M.; Gross, S. *J. Am. Chem. Soc.* 1976, vol. 98, 7169–7175.

[11] Steiger, R.; Kitzing, R.; Hagen, R.; Stoeckli-Evans, H. *J. Photogr. Sci.* 1974, vol. 22, 151–167.

[12] Zhu, X. R.; Harris, J. M. *Chem. Phys.* 1988, vol. 124, 321–332.

[13] Baumler W.; Penzkofer, A. *Chem. Phys.* 1990, vol. 140, 75–97.

[14] Tocho, J. O.; Duchowicz, R.; Scaffardi, L.; Blimes, G. M.; Dipaolo, R.; Murphy, M. *Trends Phys. Chem.* 1992, vol. 3, 31–47.

[15] Momicchioli, F.; Baraldi, I.; Berthier, G. *Chem. Phys.* 1988, vol. 123, 103–112.

[16] Murphy, S.; Sauerwein, B.; Drickamer, H. G.; Schuster, G. B. *J. Phys. Chem.* 1994, vol. 98, 13476–13480.

[17] Ponterini, G.; Momicchioli, F. *Chem. Phys.* 1991, vol. 151, 111–126.

[18] Kolesnikov, A. M.; Mikhailenko, F. A. *Russ. Chem. Rev.* 1987, vol. 56, 275–287.

[19] West, W.; Pearce, S.; Grum, F. *J. Phys. Chem.* 1967, vol. 71, 1316–1326.

[20] Dahne, S.; Leupold, D. *Phys. Chem.* 1966, vol. 70, 618–625.

[21] Karaca, S.; Elmaci, N. *Comput. Theor. Chem.* 2011, vol. 964, 160–168.

[22] Fabian, J. *Dyes and Pigments* 2010, vol. 84, 36–53.

[23] Guillaume, M.; Liergeois, V.; Champagne, B.; Zutterman, F. *Chem. Phys. Lett.* 2007, vol. 446, 165–169.

[24] Kyprianov, A. I. *Tsvet i stroyeniye tsianinovykh krasiteley* [*Color and Structure of Cyanine Dyes*]; Naukova Dumka: Kiev, 1979; 666 p.

[25] Hammer, F. M. In *The Chemistry of Heterocyclic Compounds*; Weisberger, A.; Ed.; V. 18. The cyanine dyes and related compounds; Interscience: New York, NY, 1964; 790 p.

[26] Zollinger, H.; Iqbal, A. *Color Chemistry: Syntheses, Properties, and Applications of Organic Dyes and Pigments*; 3rd revised ed.; Verlag Helvetica Chimia Acta: Zurich, 2003, 637 p.

[27] McCartin, P. J. *J. Chem. Phis.* 1965, vol. 42. 2980–2981.

[28] Khimenko, V.; Chibisov, A. K.; Gorner, H. *J. Phys. Chem. A.* 1997, vol. 101, 7304–7310.

[29] Pronkin, P. G.; Tatikolov, A. S. *High Energy Chemistry*, 2015, vol. 49, 368–371.

[30] Krieg, M.; Redmond, R. W. *Photochem. Photobiol.* 1993, vol. 57. 472–479.

[31] Caselli, M.; Latterini, L.; Ponterini, G. *Phys. Chem. Chem. Phys.* 2004, vol. 6, 3857–3863.

[32] Aramendia, P. F.; Negri, R. M.; Roman, E. S. *J. Phys. Chem.* 1994, vol. 98, 3165–3173.

[33] Rulliere, C. *Chem. Phys. Lett.* 1976, vol. 43, 303–308.

[34] Orlandi, G.; Siebrand, W. *Chem. Phys, Lett.* 1975, vol. 30, 352–354.

[35] Baumler, W., Penzkofer, A. *Chem. Phys. Lett.* 1988, vol. 150, 315–320.

[36] Rentsch, S. K. *Chem. Phys.* 1982, vol. 69, 81–87.

[37] Lee, J.; Zhu, S.-B.; Robinson, G. W. *J. Phys. Chem.* 1987, vol. 91, 4273–4277.

[38] Waldeck, D. H.; Fleming, G. R. *J. Phys. Chem.* 1981, vol. 85, 2614–2617.

[39] Akesson, E.; Sundstrom, V.; Gillbro, T. *Chem. Phys. Lett.* 1985, vol. 121, 513–522.

[40] Sundstrom, V.; Gillbro, T. *J. Phys. Chem.* 1982, vol. 86, 1788–1794.

[41] Ponterini, G.; Caselli, M. *Ber. Bunsenges. Phys. Chem.* 1992, vol. 96, 564–573.

[42] Dietzek, B.; Tarnovsky, A. N.; Yartsev, A. *Chem. Phys.* 2009, vol. 357, 54–62.

[43] Akesson, E.; Hakkarainen, A.; Laitinen, E.; Helenius, V.; Gillbro, T.; Korppi-Tommola, J.; Sundström, V. *J. Chem. Phys.* 1991, vol. 95, 6508–6523.

[44] Chibisov, A. K.; Voznyak, D. A.; Petrov, N. Kh.; Alfimov, M. V. *High Energy Chemistry*, 2009, vol. 43, 38–43.

[45] Tatikolov, A.S.; Dzhulibekov, Kh. S.; Shvedova, L. A.; Kuzmin V. A. *J. Phys. Chem.* 1995, vol. 99, 6525–6529.

[46] Tatikolov, A. S.; Akimkin, T. M.; Pronkin, P. G; Yarmoluk, S. M. *Chem. Phys. Lett.* 2013, vol. 556, 287–291.

[47] Allmann, R.; Anis, H.-J.; Benn, R.; Grahn, W.; Olejnek, S.; Waskowska, A. *Angew. Chem. Suppl.* 1983, vol. 22, 1147–1175.

[48] Pronkin, P. G.; Tatikolov, A. S.; Anikovskii, M. Yu.; Kuzmin, V. A. *High Energy Chemistry*, 2005, vol. 39, 237–243.

[49] Pronkin, P. G.; Tatikolov, A. S.; Sklyarenko, V. I.; Kuzmin, V. A. *High Energy Chemistry*, 2006, vol. 40, 252–258.

[50] Pronkin, P. G.; Tatikolov, A. S. *Spectrochim. Acta Part A,* 2018, vol. 202, 269–275.

[51] Pronkin, P. G.; Tatikolov, A. S. *High Energy Chemistry*, 2015, vol. 49, 368–371.

[52] Atabekyan, L. S.; Chibisov, A. K. *High Energy Chemistry*, 2007, vol. 41, 91–96.

[53] Tatikolov, A. S., Pronkin, P. G. *J. Appl. Spectr.* 2018, in press.

[54] Slavnova, T. D; Chibisov, A. K.; Gorner, H. *J. Phys. Chem. A* 2002, vol. 106, 10985–10990.

[55] Anikovsky, M. Y.; Tatikolov, A. S.; Kuzmin, V. A. *Int. J. Photoenergy*, 1999, vol. 1, 35–39.

[56] Anikovsky, M. Yu.; Tatikolov, A. S.; Pronkin, P. G.; Levin, P. P.; Sklyarenko, V. I.; Kuzmin, V. A. *High Energy Chemistry*, 2003, vol. 37, 398–404.

[57] Chibisov, A. K.; Gorner, H. *Chem. Phys. Lett.* 2002, vol. 357, 434–439.

[58] Valandro, S. R.; Poli, A. L.; Correia, T. F. A.; Lombardo, P. C.; Schmitt, C. C. *Langmuir* 2017, vol. 33, 891–899.

[59] Ghann, W.; Kang, H.; Emerson, E.; Oh, J.; Chavez-Gil, T.; Nesbitt, F.; Williams, R.; Uddin, J. *Inorg. Chim. Acta* 2017, vol. 467, 123–131.

[60] Wang, Z.; Chu, S.; Wang, S.; Gong, Q. *J. Chem. Phys.* 2012, vol. 137, 164502.

[61] Tahara, S.; Takeuchi, S.; Ohtani, H.; Tahara, T. *International Conference on Ultrafast Phenomena © OSA* 2016, UM2A.4.pdf. doi: 10.1364/up.2016.um2a.4.

[62] Henary, M.; Mojzych, M. *Top. Heterocycl. Chem.* 2008, vol. 14, 221–238.

[63] Panova, I. G.; Sharova, N. P.; Dmitrieva, S. B.; Poltavtseva, R. A.; Sukhikh, G. T.; Tatikolov, A. S. *Anal. Biochem.* 2007, vol. 361, 183–189.

[64] Tatikolov, A. S.; Akimkin, T. M.; Kashin, A. S.; Panova, I. G.. *High Energy Chemistry*, 2010, vol. 44, 224–227

[65] Akimkin, T. M.; Tatikolov, A. S.; Yarmoluk, S. M. *High Energy Chemistry*, 2011, vol. 45, 222–228.

[66] Bychkova, A. V.; Pronkin, P. G.; Sorokina, O. N.; Tatikolov, A. S.; Rosenfeld, M. A. *Colloid Journal*, 2014, vol. 76, 387–394.

[67] Gorobets, M. G.; Wasserman, L. A.; Vasilyeva, A. D.; Bychkova, A. V.; Pronkin, P. G.; Bugrova, A. E.; Indeykina, M. I.; Shilkina, N. G.; Konstantinova, M. L.; Kononikhin, A. S.; Nikolaev, E. N.; Rosenfeld, M. A. *Doklady Biochemistry and Biophysics*, 2017, vol. 474, 231–235.

[68] Tatikolov, A. S.; Costa, S. M. B. *Chem. Phys. Lett.*, 2007, vol. 440, 73–78.

[69] Tatikolov, A. S.; Costa, S. M. B. *J. Photochem. Photobiol. A: Chem.* 2001, vol. 140, 147–156.

In: Photoisomerization
Editor: Diego Sampedro

ISBN: 978-1-53615-313-2

Chapter 4

AZOHETEROARENE PHOTOSWITCHES: SYNTHESIS, PHOTOSWITCHING AND UTILITY

Surbhi Grewal, Debapriya Gupta, Ankit Kumar Gaur, Mayank Saraswat and Sugumar Venkataramani*

Department of Chemical Sciences,
Indian Institute of Science Education and Research (IISER) Mohali,
SAS Nagar, Knowledge City, Manauli, Punjab, India

ABSTRACT

The photoisomerization (*E-Z/Z-E*) and thermal reverse isomerization (*Z-E*) in azobenzenes are well-known phenomena that make them very useful photoswitches. For many practical applications, tuning of their photoswitching ability, and controlling the stability of *Z*-isomer are quintessential. In recent times, these aspects have been successfully addressed by the incorporation of heteroaromatic moieties in the azoarenes. The resulting azoheteroarenes have several advantages in design, synthesis, coordination and hydrogen bonding capabilities, and

* Corresponding Author's Email: sugumarv@iisermohali.ac.in.

functionalization prospects. Hence, the azoheteroarene photoswitches are expected to have several biological and medicinal implications, and also material chemistry applications. Few such applications have already been demonstrated and their growth potential is very high. At this stage, we present the historical perspectives in the use of azoheteroarenes as photoswitches, their possible classification, major synthetic strategies, salient features and controlling factors of the photophysical properties, and also selected applications through this contribution.

1. INTRODUCTION

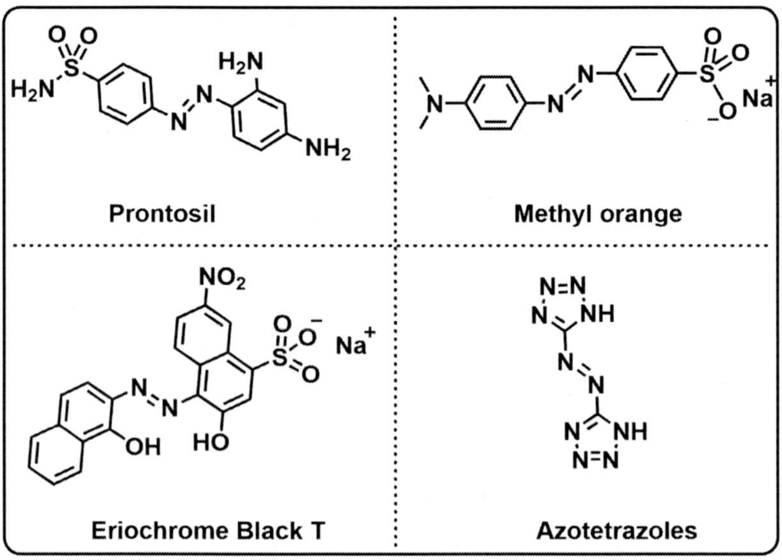

Figure 1.1. Classically important azoarene derivatives.

Azoarenes, one of the strongly colored classes of compounds is well known for more than a century. Historically, the industrial revolution in 19th century, and the need for dye molecules in the textile industries led to the introduction and studies on azoarenes [1, 2]. Later such molecules found importance in medicinal chemistry due to their inhibitory activities on bacterial pathogens. Thus, prontosil, an azo dye molecule has been utilized as an antibiotic drug in the treatment of puerperal fever [3] (Figure 1.1). However, it was later realized that prontosil was actually a prodrug that led to the formation of sulphanilamide, the actual drug [4]. Subsequently, the azoarene-based molecules, in particular with heterocyclic moieties have been studied for their medicinal activities in various therapeutic areas [5-10]. On the other hand, explosive nature has also been observed in the salts of nitrogen-rich compounds such as azotetrazoles with multiple nitrogen atoms [11]. Furthermore, the pH-induced color changes made way for few azobenzenes such as Eriochrome Black-T, methyl orange etc. to be utilized as indicators [12, 13]. The tautomerism and the associated structural and color changes were the underlying principles for their use in the volumetric titrations [14-16].

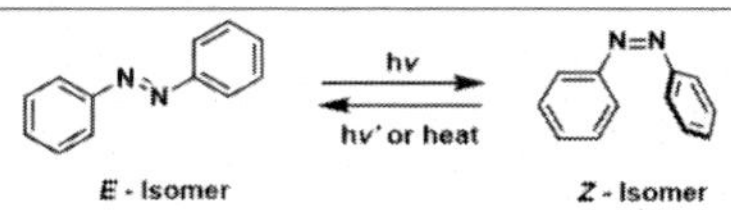

Property	*E* - Isomer	*Z* - Isomer
Dipole moment	0.5 D	3.1 D
Length	9.0 Å	5.5 Å
Molar extinction coefficient (ε)	2–3 × 10^4 M^{-1} cm^{-1}	7–10 × 10^3 M^{-1} cm^{-1}

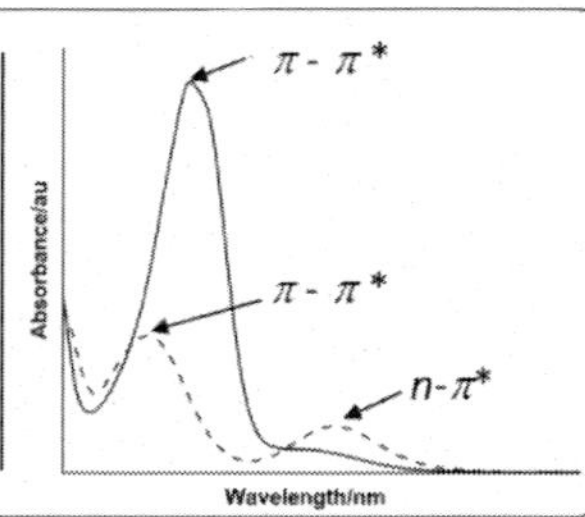

Figure 1.2. Photoisomerization in azobenzene and changes in the properties of *E*- and *Z*-isomers; Representative UV-vis spectra corresponding to *E*- (bold line) and *Z*- isomers (dotted line) of azoarene (right).

The exciting journey of azobenzenes found further twists and turns after the revelation of the *Z*-isomer for the first time by Hartley in 1937 [17]. Significantly, this report set the basis for the photoswitching ability of azobenzene. Later Fischer, Frenkel, and Wolovsky demonstrated the wavelength dependent photoisomerization equilibria in azobenzenes [18]. Following the footprints of such impeccable works, photoisomerization between the *E*- and *Z*-isomers of several azoarenes has been studied with great interest. The salient features of these two isomers are the variation in their physical properties such as dipole moment, absorption spectral features, planarity, π-conjugation etc. that can be followed in understanding the isomerization processes [19, 20]. Typically, the azo group exhibits two electronic absorption features namely, π–π* and n–π* bands (Figure 1.2). Due to the planarity, the *E*-isomer exhibits a strong π–π* (symmetry allowed transition) and a weak n–π* (symmetry forbidden transition) absorption features. Upon *E*-*Z* isomerization, and consequent breaking of planarity and symmetry, the azo group shows a weak π–π* band often with a blue shift. On the other hand, the n–π* band appears with a characteristic enhancement in the intensity.

The nature and separation between the π–π* and n–π* absorption features for both *E*- and *Z*-isomers depend on the substituents, as a consequence, the type of substituents can dictate the switching ability. [20, 21] The forward isomerization step invariably involves a photochemical excitation. Notwithstanding the *Z*-isomer formation through photoisomerization channel, the reversal of it to attain the thermodynamically stable *E*-isomer can happen either through light (photochemically) or under dark (thermally). In short, the *Z*-isomer demands wavelength dependent photoisomerization for the formation, and also possesses a variable thermal stability even under dark condition. Several spectroscopic and computational investigations have been explored in understanding the forward and reverse isomerization phenomena. (see section 5) These included mechanistic aspects related to the isomerization, photoswitching ability and thermal stability of *Z*-isomer. For comparison of various azoarenes, the following factors are considered to be important, and frequently used: (a) quantum yield (ϕ, a quantity deciding the efficiency of the isomerization reaction at a particular wavelength of light with known intensity), (b) photostationary state (PSS, an attainment of steady-state equilibrium upon light-induced isomerization between *E*- and *Z*-isomers, where the light has been absorbed by at least one of the components; At PSS the rates of forward and reverse isomerization steps are equal for each of the transient molecular entities formed), and (c) half-life ($\tau_{1/2}$, a measure of the lifetime of the less stable isomer through a thermal reverse isomerization process under dark). Often such properties are dependent on the conditions such as solvent, temperature, concentration etc.

As a consequence of photoswitching behavior, azoarenes attracted several experimentalists. Over several decades, various synthetic strategies have also been developed to access a wide range of azoarenes with ease using cheaper starting materials and milder conditions [19, 22]. Furthermore, the efficient and robust photoswitching nature of azoarenes made way for their potential applications in a variety of fields such as molecular machines [23-25], material chemistry [26-30], biomolecular [31-40] and medicinal chemistry [41-45] including photopharmacology [46-48]. For many aforesaid applications, controlling and tuning of photoswitching behavior and the stability of *Z*-isomer became very crucial. In this regard, structural modifications and an introduction of substituents have been considered. Investigations on both symmetrically and unsymmetrically substituted azobenzenes, and also the heterocyclic-based azoarenes have greatly influenced the progress of azoarene chemistry [49, 50]. In recent times, the azoheteroarenes gained greater importance over simple azobenzenes due to better and quantitative photoswitching ability, the long half-life of *Z*-isomer, tunability and functionalization prospects. However, to the best of our knowledge, there is no detailed review on azoheteroarenes.

At this juncture, a review of the progress in the chemistry of azoheteroarenes, and further challenges will be enormously helpful. This chapter will mainly focus on the historical survey, synthetic methods to access various azoheteroarenes, their photoswitching aspects, their functional utility, and future prospects. We mainly considered the molecules with at least one heteroatom-containing ring as a part of the azoarene systems. In this regard, we have classified them to accommodate different types of azoheteroarenes. Our major intention was to report various azoheteroarenes exhibiting photoisomerization. Still, we have cited several works of literature based on azoheteroarenes, where the primary focus was not photoisomerization processes, in order to consider them in future as photoswitches or for relevant applications.

2. Historical Survey

Historically, the journey in search of new azo dyes made way for the access to several azoheteroarenes [51-99]. Besides that, the understanding of *E-Z* photoisomerization in azobenzenes led to the photoswitching studies of azoheteroarenes. In 1951, Le Favre demonstrated the light-induced isomerization of 2, 2'-azopyridine [100]. Following this work, Campbell et al. have reported the geometrical isomers of various symmetrical and unsymmetrical azopyridines [101]. In order to understand the structural aspects of both the isomers, the changes in their dipole moments, melting points and UV-vis absorption properties have been envisaged. In order to understand the status and progress of the azoheteroarenes, we searched for literature using SciFinder® with various keywords such as "azoheteroarene", "azopyridine", "azoimidazole" etc. (by varying the name of the heterocycles). The survey unveils the tremendous increase in the publications on azoheteroarenes in recent decades despite the fact that such molecules were known for more than hundred years (Figure 2.1). However, refinement of those searches with keywords such as "photoswitching", "isomerization" etc. restricted the number of publications that too mainly appear from the past decade.

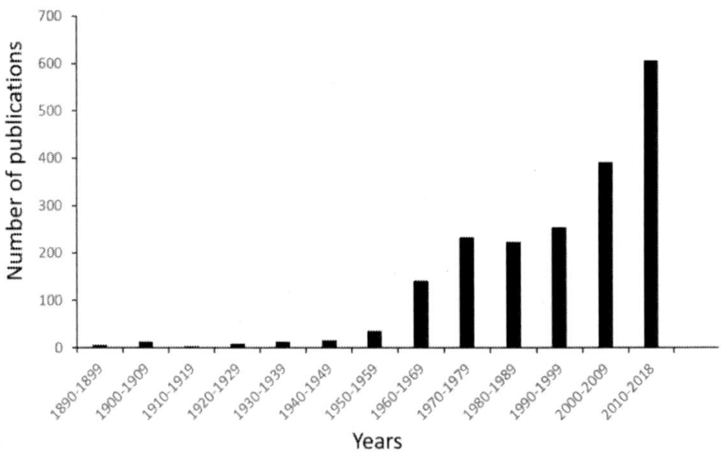

Figure 2.1. Decade-wise reports on various azoheteroarenes (Source: SciFinder®, searched in October 2018).

Among the various azoheteroarenes, azopyridines are one of the widely studied analogs. Even after the revelation of azopyridine photoswitching, indeed there was a quiet period with respect to the photoisomerization aspects of other heterocyclic analogs. Except the initial reports on azopyridines, most of the other reported azoheteroarenes have been synthesized mainly as therapeutic targets [102-104] in medicinal chemistry, and as colorants in dye industry [105, 106] until recently. In this regard, several molecules with a wide range of heterocyclic structures have been synthesized and studied for their therapeutic effects as anti-fungal, antibiotics etc. in the medicinal chemistry [107-111]. Simultaneously, various synthetic strategies have also been developed in obtaining many azoheteroarenes. Moreover, the utility of common intermediates such as "β-diketohydrazone derivatives" diversified the access to different azoheteroarenes [112-115]. However, there has been a new development in using azoheteroarenes in different fields such as: a) directing groups for synthesis [116], b) heterogeneous reagent [117], and c) non-linear optics [118] etc. Interestingly, the structural variation, in particular, the introduction of heteroatoms provides support to the coordination [119-125], hydrogen bonding prospects, and other non-covalent interactions [126-137] that extend the applications. Based on these prospects, they have been utilized for many other applications such as coordinating ligands [123, 138, 139], co-crystals with thermo- and photoresponsive behavior [126, 127, 140], as a linker in metal-organic frameworks (MOFs) [141-144] etc. Obviously, the photoswitching behavior of azopyridines was also fully utilized in obtaining applications such as photoswitchable contrast agents [145-147] and photoswitchable metabotropic glutamate receptors [148] etc.

The heterocyclic analogs of the azoarenes acquired an important status in recent times. Attempts have been made towards achieving the important goals in improving the photoswitching behaviour, and bringing tunability as well as stability to the azoheteroarenes [49, 50, 149-153]. Prior to this situation, major attention was given to the detailed investigations on the kinetics and mechanistic aspects of azo coupling reactions in the formation of several azoheteroarenes [154-160], whereas, their photoswitching behavior has been seldom reported. The role of azoheteroarenes with photoswitching applications attained a new era after the reports on photoswitchable contrast agents reported by Herges group [147]. By using nickel porphyrin and the azopyridine, they were able to modulate the magnetic properties by light [145, 146, 161]. As an extension, they have also developed a synthetic strategy to access phenylazoimidazole [49, 162]. The chemistry of azoheteroarene-based photoswitches attained a new era after the incorporation of imidazole in the azoarene moiety. Indeed, the introduction of five-membered heterocyclic systems tremendously influenced the *Z*-isomer stability and also photoswitching efficiency. Following the footsteps of this work, Fuchter and coworkers further improved the *Z*-isomer stability by replacing the imidazole with *N*-methylpyrazole unit [150]. In yet another extensive work, they utilized different five-membered heterocyclic based azoheteroarenes and demonstrated that the *N*-methylation is crucial, whereas the *ortho* substitution decreases the stability of *Z*-isomer in azopyrazoles [150]. Quite recently, König and coworkers have developed azoindoles that can be tunable between nanoseconds to days [149]. Apart from the steric and electronic effects through substituents, the hydrogen bonding is equally important in controlling the stability of the *Z*-isomer in arylazopyrazoles [151]. Following the recent trend in visible light-induced molecular photoswitching, Kolarski et al. have synthesized azoheteroarenes based on the purine scaffold for photopharmacological applications [163]. All these results support the possibility of using azoheteroarene as better photoswitches in futuristic applications. Now the stage is reached for several novel applications and improvement in the existing systems.

3. Classification(s) and Types of Azoheteroarenes

3.1. General Classification

There are many heterocyclic-based azoarenes available in the literature. Due to the difficulty in organizing them, we introduce a general classification scheme. In a broader perspective, we classified all the azoheteroarenes into the following four different types based on the number and nature of heterocycles (Figure 3.1):

A. Class–I (Monoheteroaromatic ring containing azoheteroarenes)
B. Class–II (Bisheteroaromatic rings containing azoheteroarenes)
C. Class–III (Fused heteroaromatic ring(s) containing azohetero-arenes)
D. Class–IV (Partially or fully saturated heterocyclic ring containing/miscellaneous azoheteroarenes)

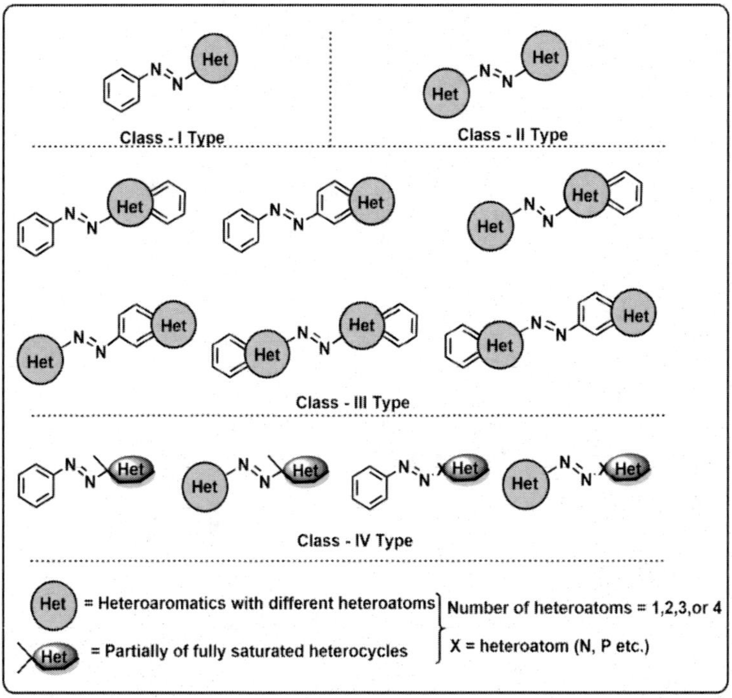

Figure 3.1. General classification of azoheteroarenes.

3.2. Class–I Azoheteroarenes

This is the most frequently reported class of azoheteroarenes. Overall, this class of azoheteroarenes contains one heterocyclic ring, which is directly attached to the azo group (Figure 3.2). Simple or substituted phenyl group used to be the attachment at the other side of the azo group. This class of azoheteroarenes can be further divided into sub-classes based on the ring size of the heterocycle, and the number of heteroatoms. Mono-azoheteroarenes can be divided into two sub-types, based on the ring size, i.e., 5-membered heterocyclic rings and 6-membered heterocyclic rings. Further, these can be classified on the basis of the number of heteroatoms on heterocyclic ring as mono-, di-, tri- and tetra-heteroatom containing rings.

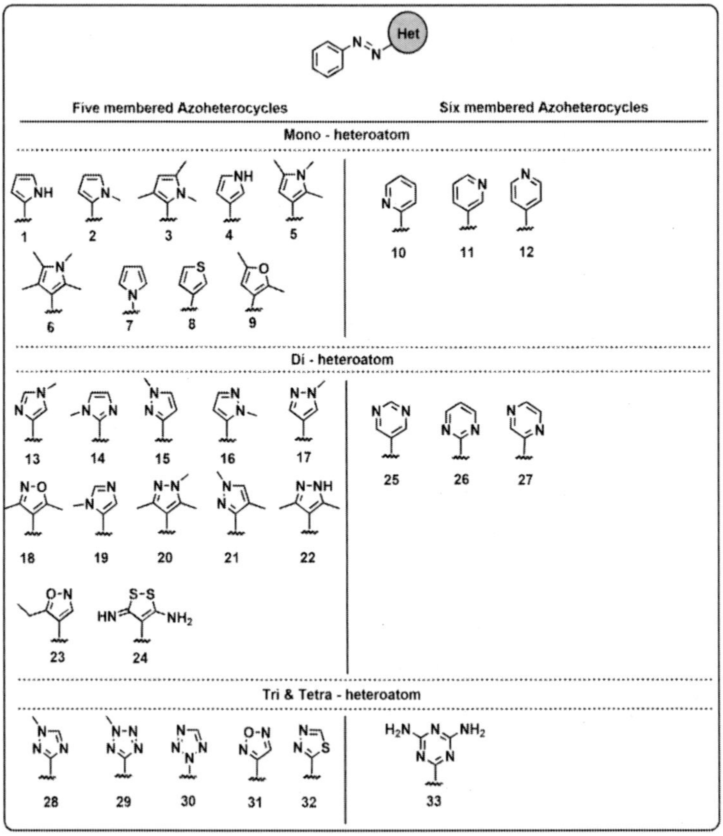

Figure 3.2. Class-I type azoheteroarenes: Possible five-membered and six-membered monoheterocyclic azoheteroarene (Compounds and references: 1 [164, 165], 2 & 3 [50, 150, 166], 4 [164, 167], 5 [150], 6 [150], 7 [168], 8 [169], 9 [170], 10 [101, 171-174], 11 [153, 101, 175-177], 12 [124, 101, 178-180], 13 [49, 150], 14 [150, 181, 182], 15 [150], 16 [150], 17 [50, 150, 166, 183], 18 [184-186], 19 [49, 150, 187], 20 [50, 150, 151, 183], 21 [150], 22 [125, 150, 151, 188], 23 [189], 24 [190], 25 [153, 191], 26 [192, 193], 27 [153], 28 [150], 29 [150], 30 [194], 31 [195], 32 [196], 33 [197]).

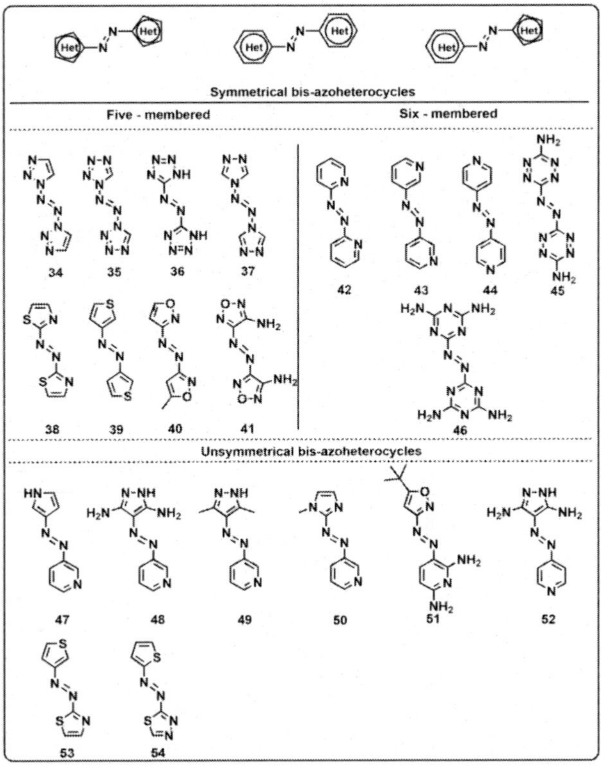

Figure 3.3. Class-II type azoheteroarenes: Possible symmetrical, unsymmetrical with five-/six-membered and mixed type bis-heterocyclic azoheteroarenes (Compounds and references: 34 [198, 199], 35 [8, 200, 201], 36 [202, 203], 37 [204-208], 38 [117, 209-211], 39 [212], 40 [117, 213-215], 41 [216-220], 42 [100, 175, 221-224], 43 [101, 225-229], 44 [140, 230-233], 45 [234-237], 46 [238-240], 47 [167], 48 [241], 49 [152], 50 [242, 243], 51 [244], 52 [245-247], 53 [195], 54 [195]).

3.3. Class–II Azoheteroarenes

The second important class of azoheteroarenes is the one having two heteroaromatics in either side of the azo group (Figure 3.3). Indeed, both the heterocycles are directly attached to the azo group. Furthermore, these systems can be either symmetrically or unsymmetrically connected with respect to the azo group. Likewise, in the previous class, these can also be sub-divided based on the size of the ring (5 or 6 members) and the number of heteroatoms.

3.4. Class–III Azoheteroarenes

In this class, one or two fused heterocycles will be directly attached to the azo group (Figure 3.4). These can once again be symmetrical or unsymmetrical with respect to the azo group. Besides that, the fused heterocyclic can be connected either at the heterocyclic part or through the benzene (non-heterocyclic) part. Furthermore, they can be classified as systems with five-membered or six-membered fused heterocyclic rings.

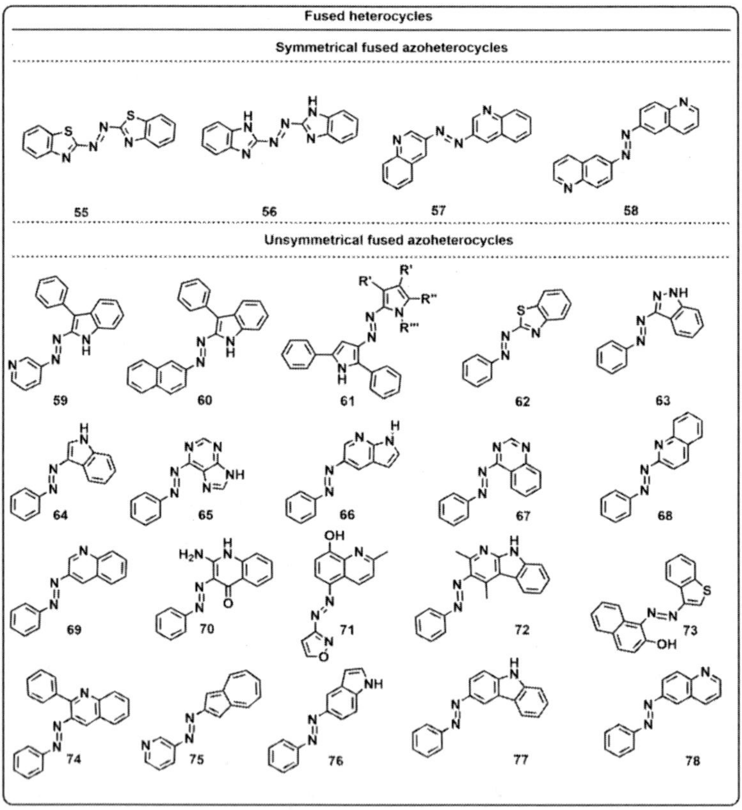

Figure 3.4. Class-III type azoheteroarenes: Possible symmetrical, unsymmetrical with five-/ six-membered and mixed type fused heterocyclic based azoheteroarenes (Compounds and references: 55 [117, 213, 248-253], 56 [117, 209, 254, 255], 57 [225, 256], 58 [257], 59 [258, 259], 60 [258], 61 [260], 62 [261, 262], 63 [153, 263], 64 [264], 65 [153], 66 [153], 67 [153], 68 [265, 266], 69 [153, 267, 268], 70 [269], 71 [270], 72 [271], 73 [169, 272], 74 [273], 75 [274], 76 [153, 275], 77 [153, 276, 277], 78 [278]).

3.5. Class–IV Azoheteroarenes

In this class, heterocycles with fully or partially saturated rings are connected to the azo group (Figure 3.5). More examples are available in which the azo group is connected to the ring nitrogen atom forming a triazene link. Several miscellaneous types of azoheteroarenes, which cannot be classified under other three categories will also be accommodated in this class.

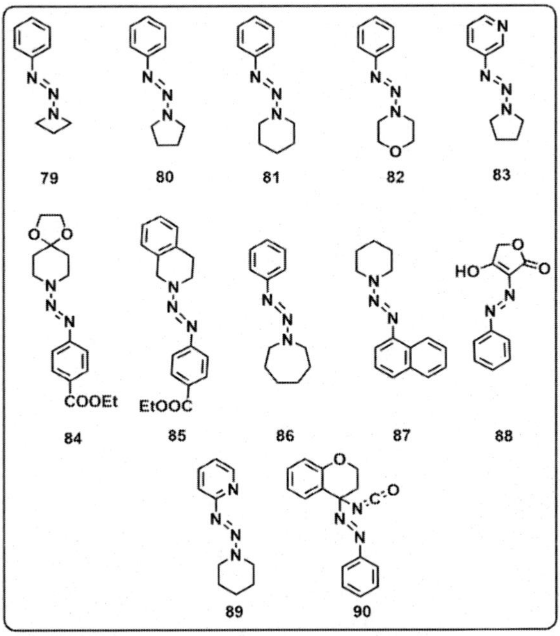

Figure 3.5. Class-IV type azoheteroarenes: Fully or partially saturated heterocyclic based and miscellaneous azoheteroarenes (Compounds and references: 79 [279], 80 [258, 279-281], 81 [279], 82 [279], 83 [258, 281-283], 84 [284], 85 [284], 86 [279], 87 [280], 88 [285, 286], 89 [287], 90 [288, 289]).

4. Synthetic Strategies

Herein, we briefly discuss the popular synthetic strategies that have been adopted to access various azoheteroarenes. In the first part of this section, the most common and classical methods will be discussed. In the later part, a few representative modern approaches to accessing complex azoheteroarenes will be given. The synthesis of azoheteroarenes mostly follows the same synthetic methods as azobenzenes with a few exceptions. The most commonly used synthetic methods are as follows:

4.1. Azo coupling reactions and reactions involving diazonium salts
4.2. Mill's method
4.3. Oxidative coupling of heterocyclic amines
4.4. Reductive coupling of nitroarenes
4.5. Synthesis and dehydrogenation of arylhydrazines
4.6. Recent approaches to accessing azoheteroarenes
4.7. Miscellaneous synthetic methods

4.1. Azo Coupling and Reactions Involving Diazonium Salts

Azo coupling is one of the classical methods for the synthesis of azobenzenes [19, 22, 290-292], which involves the *in situ* generations of diazonium salt using a primary aromatic amine, and consequently, reacting it with electron-rich aromatic nucleophiles (Figure 4.1a). Since several heterocyclic diazonium salts can be readily accessible [293], this reaction can be extended to synthesize azoheteroarenes, particularly in the synthesis of class-I type azoheteroarenes. Alternatively, electron-rich heterocycles can also be used as nucleophiles to access class-II type azoheteroarenes. Typically, this reaction leads to either *para* or *ortho* derivative (if *para* position is blocked) with respect to the electron-donating group present in the electron-rich aromatic nucleophile. The commonly used aromatic nucleophiles include anisole, β-naphthol, phenol, *N*-methylaniline [169, 272, 274, 294-299], and even calix[n]arene [300-302] etc. or electron-rich aromatic heterocycles like 2, 6-diaminopyridine [244, 257, 260, 270, 303]. Similarly, diazonium salts of aniline and its derivatives can also be reacted with heteroaromatic compounds like *N*-methylpyrrole, indole etc. or with non-aromatic heterocyclic compounds like tetrahydropyrrole [49, 149, 150,

167, 170, 280, 284, 286, 304-314] to obtain class-I and class-IV type azoheteroarenes, respectively (Figure 4.1b).

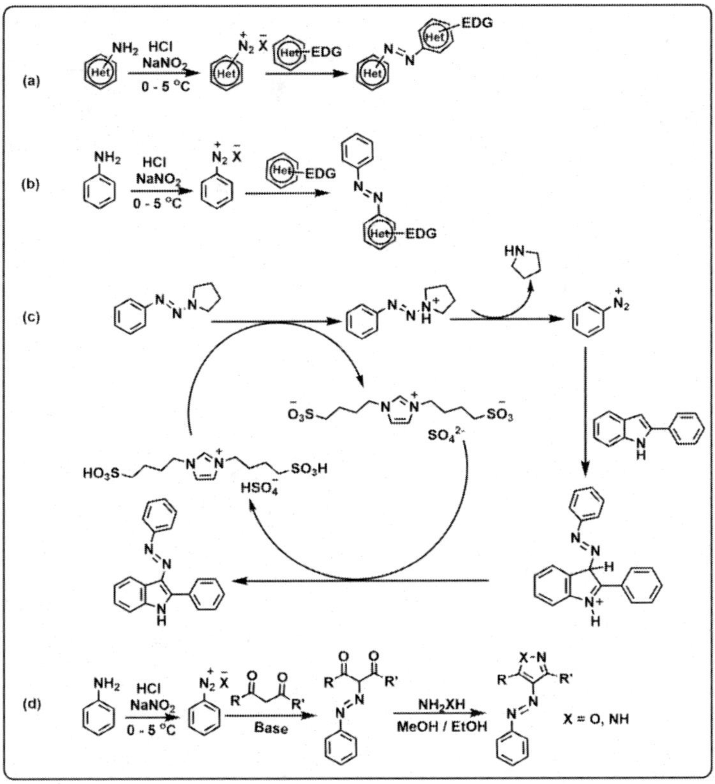

Figure 4.1. Common synthetic strategies to access azoheteroarenes: (a) Azo coupling (for class-II type azoheteroarenes); (b) Azo coupling (for a class-I type azoheteroarenes) (c) Ionic liquid promoted diazenylation with masked diazonium ion equivalent. Reproduced with permission from ref. 258; Copyright 2016 American Chemical Society. (d) Reaction with active methylene compounds.

In the majority of the cases, nitrous acid (generated *in situ* using sodium nitrite in the presence of an acid), is utilized for diazotization [315]. Various acids such as HCl, HBr, HBF_4, PTSA, H_3PO_4 can also be used depending on the stability of the diazonium salt, which depends on the counter anion, temperature etc. Reagents such as nitrosonium tetrafluoroborate, *tert*-butyl nitrite or isopentyl nitrite can also be used to prepare the diazonium salts from aromatic amines under non-aqueous conditions [169]. Thus, the aromatic nucleophile present in the medium can undergo electrophilic aromatic substitution reaction with the diazonium salt leading to the azo compound.

Alternatively, aryltriazines have also been used as a masked diazonium ion equivalent in Lewis acid promoted chemoselective aryl diazenylation of β-dicarbonyls [316]. In 2016, Cao et al. reported the Bronsted acidic ionic liquid promoted diazenylation of heterocyclic aromatic compounds with aryltriazene as a masked diazonium ion equivalent [258] (Fig 4.1c). Upon activation of the aryltriazene, a tetrahydropyrrolyl cation can be generated, which in turn leads to the diazonium ion that participates in the aromatic electrophilic substitution with 2-phenylindole or other derivatives.

Unlike the classical azo coupling reactions, which demands an activated aromatic nucleophile, Severin's group developed a strategy for unactivated aromatic systems. In this strategy, Lewis acid induced C-H activation process has been utilized to generate highly reactive dicationic diazonium salts that do not require an activated aromatic nucleophile for the azo coupling [317]. This strategy can be applicable to unactivated (benzene, mesitylene), deactivated (C_6H_5F) and heteroaromatic arenes, apart from that, polystyrene can also be used as a coupling partner. The plausible mechanism involves an initial complex between $AlCl_3$ and NHC-N_2O adducts that abstracts oxygen, resulting in the formation of a dicationic diazonium salt, which undergoes a reaction with the arene to obtain class-I and II type azoheteroarenes. The same group has also demonstrated that N_2O adducts of mesoionic carbenes can also be used as a precursor for azoimidazolium salts [318].

Since the diazonium ion is a weak electrophile, in cases where the nucleophilic coupling partner is not activated, bases such as sodium acetate, potassium carbonate etc. can be used to increase the nucleophilicity of the coupling partner. Using a strong base like *n*-BuLi, the azo coupling can be performed at the pyrrole nitrogen [168]. Compounds having active methylene groups such as acetylacetone, malononitrile, and 3, 5-heptanedione can also be coupled with the weakly electrophilic diazonium salt in basic medium. The resulting arylhydrazones can further be reacted with hydrazine, hydroxylamine etc. to generate 3, 5-disubstituted arylazopyrazole, pyrazolone, and isoxazole based heteroazoarene derivatives, respectively [151, 152, 319-325] (Figure 4.1d). Using this method, unsymmetrically 3, 5-substituted class-I azoheteroarenes can be prepared. In the case of 3, 5-unsubstituted pyrazole derivative, enamine (*E*)-3-(dimethylamino)-2-propenal has to be used as a starting material [321].

In a recent example, Li et al. developed a base-promoted decarboxylative azo coupling strategy involving aryldiazonium salts with carboranyl carboxylic acids for the synthesis of *trans*-azocarboranes [304]. A range of diazonium salts including heterocyclic diazonium salts has been shown to be compatible with this method. Unlike other decarboxylative procedures, this does not proceed through a radical pathway and instead, it is quite similar to the coupling of organometallics RM [M = MgX, ZnX, Li; X = Cl, Br] with diazonium salts for the synthesis of azoarenes [305-308]. Arylcyclohexyltriazenes (class-IV type azoheteroarenes) can be synthesized by using phenylmagnesium bromide as *C*-nucleophile with a lithium alkylamide-N_2O adducts [357].

4.2. Mill's Method

The reaction of aromatic nitroso derivatives with aniline or other aromatic amines is known as Mills reaction. Mills reaction is one of the classical methods for the synthesis of azoarenes [19, 22]. This reaction is usually carried out in an acidic medium (usually in acetic acid) [332-333]. Mechanistically, in acidic condition, the aromatic amine as a nucleophile attacks at the electrophilic *N*-center of the nitroso compound. A subsequent proton shift and acid-mediated dehydration lead to the formation of the azo compound [334] (Figure 4.2a). In contrast, for generating azoheteroarenes such as azopyridines using deactivated aminopyridines, basic medium is necessary [50, 102, 150, 178, 212, 261, 267, 335-337]. Here, the base-assisted deprotonation from the amines is the key step, and the resulting *N*-centered anion attacks the nitrosoarene, and a subsequent dehydration step leads to the formation of azo compound (Figure 4.2b). The nitrosoarenes can be prepared from their corresponding aniline derivative using $FeCl_3$ [212],

Oxone® (biphasic heterogeneous system $2KHSO_5.KHSO_4.K_2SO_4$) [339, 340], and several other oxidizing agents [19, 22]. Alternatively, the nitroso compounds can be generated from their corresponding nitro compounds over two steps, viz, a partial reduction using Zn/NH_4Cl, followed by an oxidation step using $FeCl_3$ solution [145].

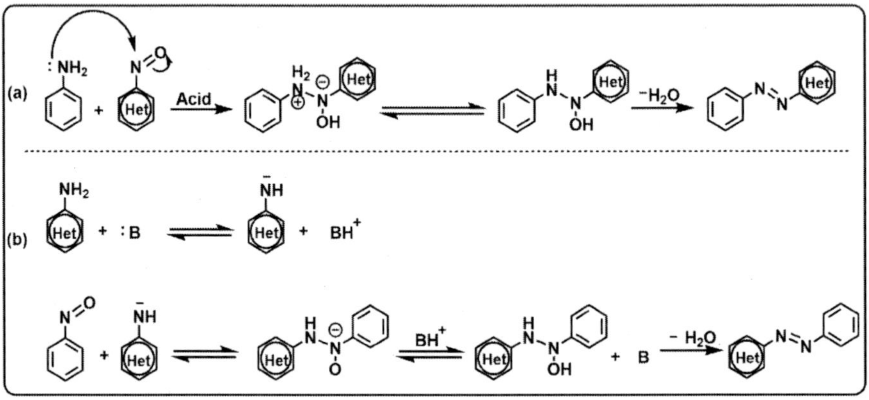

Figure 4.2. Mills method: (a) in an acidic medium. Reproduced with permission from ref. 22; Copyright 2011 Royal Society of Chemistry; and (b) in basic medium. Reproduced with permission from ref. 338; Copyright 1953 Royal Society of Chemistry.

4.3. Oxidative Coupling of Heterocyclic Amines

Oxidative coupling of aromatic heterocyclic amines is one of the popular methods for the synthesis of symmetrical azoheteroarenes of types class-II and III (Figure 4.3a). As early as 1898, Thiele et al. reported the synthesis of 5, 5'-azotetrazolate as salts of metal cations (K^+, Ca^{2+}, Ba^{2+}, Ag^+, Hg^+, Pb^{2+}) and protonated nitrogen bases by oxidizing of aminotetrazole using potassium permanganate [11, 341-348]. 3-Aminopyrazoles can also be oxidized by manganese dioxide to form azobispyrazole [349]. Sodium dichloroisocyanurate (SDCI) in acetic acid or acetonitrile can also be used for the oxidation of heterocyclic amines [198, 350]. Other oxidative reagents such as $KMnO_4$ in HCl [217, 351], saturated aq. Br_2 solution [352], and Cl_2 solution in biphasic chloroform/

aqueous sodium bicarbonate solution [353], $PhI(DMAP)_2(OTf)_2$ [354], cetyltrimethylammonium dichromate [253], and CuBr in the presence of NMO [225], NaOCl [57, 355], and $PhI(OAc)_2$ [209], $TiCl_3$-HBr and NH_4VO_3 [356], *t*-BuOK in presence of NH_3 [215] can also be used. Recently Dutta et al. have used an inverse-micelle-templated mesoporous manganese oxide materials in oxidizing 2-aminopyridine to 2, 2'-azopyridine [357]. The mechanism of oxidative couplings depends on the reagents used, however, a single electron transfer (SET) step invariably plays an important role. For instance, Singh and coworkers have proposed a plausible SET based mechanism for the CuBr-NMO oxidative system [225] (Figure 4.3a). The initial SET from Cu(I) to NMO leads to the generation of aminyl radical and Cu(II) salt through homolytic cleavage of N-O bond. The excess Cu(I) present in the medium transfers a single electron to the aminyl radical to give *N*-methylmorpholine and Cu(II) salt. Now the Cu(II) salt coordinates with aniline to form a complex, which forms a radical cation through a SET mechanism. Dimerization of the radical cations and consequent dehydrogenation affords the corresponding azobenzene.

4.4. Reductive Coupling of Nitroarenes

In 1986, while investigating the reduction of the nitroimidazole drug misonidazole, Gattavecchia et al. synthesized a class-II type symmetric azobis(2-imidazole). In this regard, they have utilized nitroimidazole and Zn dust in an aqueous NH_4Cl solution [358]. In their synthesis of photoswitchable bases, Weston et al. adopted the same strategy for obtaining *N*-methylated azobisimidazoles [359]. The plausible mechanism involves the formation of nitroso and hydroxylamine derivatives as intermediates (Figure 4.3b). Both of them eventually form their corresponding radical anion, followed by dimerization. A subsequent dehydration step leads to the formation of an azoxy compound in both acidic and basic medium. Finally, a deoxygenation step leads to the heteroazoarene [360-361]. It is noteworthy to mention that Gevorgyan et al. adopted silicon-mediated water splitting reaction for the formation of class-I type azoheteroarenes from corresponding nitro compounds [278].

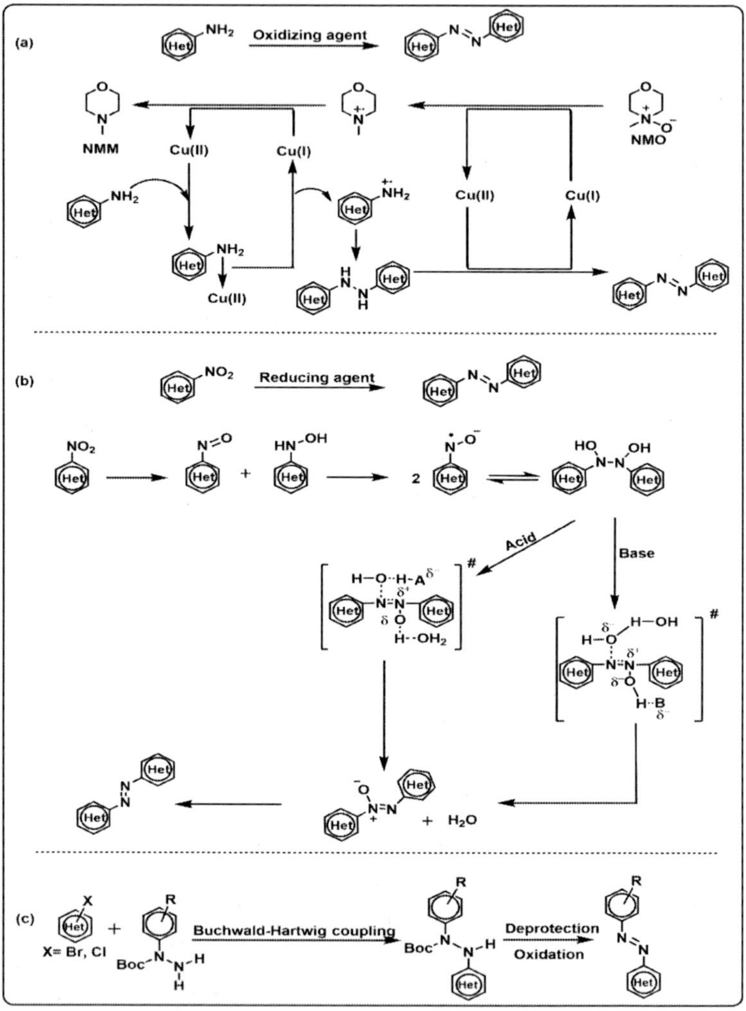

Figure 4.3. (a) Oxidative coupling (a brief mechanism has been included for CuBr-NMO oxidative system; Reproduced with permission from ref. 225; Copyright 2015 Royal Society of Chemistry; (b) Reductive coupling (with general mechanism); Reproduced with permission from ref. 22; Copyright 2011 Royal Society of Chemistry and (c) Dehydrogenation of arylhydrazines.

4.5. Dehydrogenation of Arylhydrazines

Azo group can also be introduced by dehydrogenation of corresponding *N*, *N*'-diarylhydrazines. Commercially available halo heteroarenes can undergo either an uncatalyzed nucleophilic aromatic substitution with arylhydrazines [163] or palladium catalyzed Buchwald-Hartwig coupling with Boc-protected arylhydrazines [153] to produce the corresponding diarylhydrazines in good to excellent yields. Finally, oxidants such as *N*-bromosuccinimide in acetonitrile or Cl_2 in H_2O and $CHCl_3$ [205, 234, 238, 240, 353, 362, 3634], CuI [364] or under O_2 [163] are known to convert hydrazine (-NH-NH-) to the azo (-N = N) group. Adopting a similar procedure, Garcia—Amoros et al. condensed *p*-benzoquinone with 2-hydrazinopyrimidine, followed by *in situ* dehydrogenation using perchloric acid to synthesize 2-(4-hydroxyphenylazo) pyrimidine, a class-I type azoheteroarene [365].

4.6. Recent Approaches to Accessing Azoheteroarenes

4.6.1. Functionalization via S_N^{Ar} Substitution

Recently, Puente et al. described the synthesis of arylazoindazole photoswitches via base-induced nucleophilic aromatic substitution of formazan [366] (Figure 4.4a). The resultant class-III type asymmetric arylazoindazole can be further functionalized by highly selective nucleophilic displacement reactions. The Formazan can be synthesized by standard diazonium coupling of imine derivative of 2, 3, 4, 5, 6-pentafluorobenzaldehyde and phenylhydrazine [367].

4.6.2. Diazenylation with Arylhydrazine Hydrochlorides

Barak et al. have utilized basic and aerobic conditions in diazenylation of active methylene and *N*-heterocyclic compounds using arylhydrazine hydrochlorides [368] (Figure 4.4b). It is well-known in the literature that phenyl radical can be generated by oxidizing phenylhydrazine through an unstable species phenyldiazene as an intermediate [369-372]. Although earlier literature reported carbodiazenylation of olefins by using phenylhydrazine as the aryldiazene source [373], the utilization of air sensitive arylhydrazines remained a challenge. In this work, they used the air-stable arylhydrazine hydrochloride, which was oxidized to an aryldiazene cation radical under photoredox condition. The iodine mediated generation of aryldiazene radical intermediate can be scavenged by either active methylene compound or indole leading to the formation of class-III type unsymmetrical azoheteroarenes. However, one important prerequisite in this protocol is that the active methylene compounds should contain at least one nitrile group.

4.6.3. Cascade Reaction Procedures

In 2015, Pandit et al. reported a silver catalyzed cascade reaction for synthesis of multifunctionalized azopyrazoles. The α-diazo-β-keto ester first undergoes a condensation reaction with arylhydrazine leading to *N*-substituted pyrazole ring formation (Figure 4.4c). The Ag(I) catalyst present in the reaction initiates the loss of nitrogen in the diazo component resulting in a metal-carbenoid species, which undergoes N-H insertion with the excess arylhydrazine present in the medium, followed by oxidation leading to class-I type azoheteroarene [374]. In 2016, the same group utilized iodine catalyzed one-pot procedure to install two different hydrazine moieties in the same molecule [375]. The mechanism was predicted to be quite similar to the previous reaction. The only exception, in this case, is the participation of α-iodo-β-keto ester as an intermediate, which involves in the pyrazole ring formation and a subsequent substitution by an *N*-nucleophile (through Ag(I) aided removal of iodo component). Finally, oxidation and enolization steps lead to the generation of the class-I type azoheteroarene.

4.6.4. Symmetrical Azoheteroarenes from Aryl Azides

Uyeda et al. have reported the catalytic dimerization of nitrene to generate the symmetrical azoheteroarene (Figure 4.4d). In this regard, they have utilized a dinuclear nickel complex as a catalyst. Interestingly, the *in situ* generated nitrene from its azide precursor undergoes dimerization with the evaluation of nitrogen. Based on the catalyst design, the steric factor in the ligand, and the dinuclear nickel connection lead to the formation of *Z*-isomer, which was weakly coordinated with the catalyst and so, effectively avoided the byproduct formation. Apart from that, the catalyst was found to tolerate a wide range of functional groups [376].

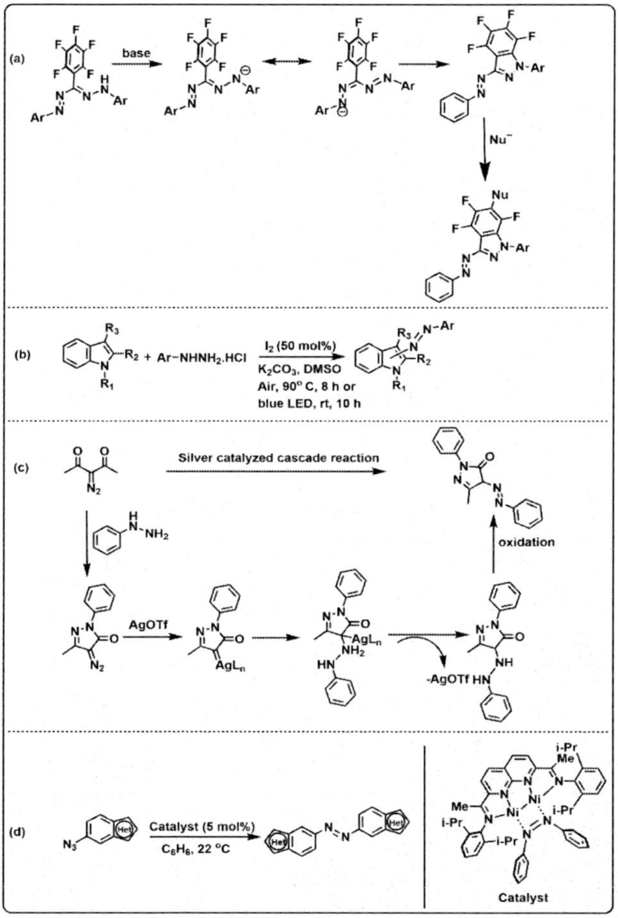

Figure 4.4. (a) Functionalization via S_N^{Ar} substitution. Reproduced from ref. 366; Copyright 2017 American Chemical Society. (b) Diazenylation with arylhydrazine hydrochlorides. Reproduced with permission from ref. 368; Copyright 2018 American Chemical Society. (c) Cascade reaction through pyrazole formation, N-H insertion, and oxidation sequence. Reproduced with permission from ref. 374; Copyright 2015 John Wiley & Sons. (d) Dimerization of nitrenes. Reproduced with permission from ref. 376; Copyright 2018 American Chemical Society.

4.7. Miscellaneous Methods

In 1967, Partridge et al. reported that with hydrazine hydrate, pyrazolotriazene underwent ring fission to give pyrazol-3-ylazopyrazole [377]. Other reported procedures include photolysis of *N*-(benzotriazol-1-yl-phenylmethylene)-*N′*-arylhydrazines for the synthesis of phenanthridin-6-yl-2-phenyldiazine [273], and the synthesis of 3-azoquinoline derivatives via a two-step procedure including Mannich reaction of α-ketohydrazones to 2-acetamidobenzaldehyde and subsequent elimination of *N*-benzylpiperazine from the Mannich adducts [378]. For the synthesis of arylazoimidazolium dyes, oxidative coupling of hydrazones with activated arenes has been reported [379-381].

5. Photoswitching Behavior

5.1. General Aspects on Photoswitching

The reversible configurational changes about the azo group upon irradiation with the light of appropriate wavelength play a crucial role in utilizing azoarenes as photoswitches. The variation in geometries and the accompanying changes in the physical properties of *E*- (*trans*) and *Z*-isomers (*cis*) are important for many applications. Generally, the planar *E*-isomer is thermodynamically more stable compared to its non-planar *Z*-isomer, however, certain bridged azobenzene and azobenzenophane have been reported to be in the *Z*-isomer as their more stable form due to ring strain [382, 383].

Prior to the discussion on photoswitching behavior and the thermal stability aspects of the *Z*-isomer of various azoheteroarenes, it is very important to understand those aspects in azobenzene. Such an introduction is necessary to understand the effects of heterocycles in the photoisomerization processes and *Z*-isomer stability. For understanding those aspects in azobenzenes, several studies related to the excited states, their properties including lifetimes, substituent effects, solvent effects, and also isomerization mechanisms etc. have been carried out. In this regard, a wide range of techniques including time-resolved spectroscopy and computations have been utilized. The most significant results obtained based on those studies are as follows:

5.1.1. Absorption and Photoisomerization

The absorption properties of the native *E*-form of the azobenzenes often found to be a key aspect in understanding the photoswitching. The extent of separation between the π–π^* and n–π^* absorbance bands can be used to classify azobenzenes in to different groups as follows: (i) Azobenzenes type with well-separated π–π^* and n–π^* absorbance bands (Typically this class have a long lifetime for the *Z*-isomer); (ii) Aminoazobenzenes type with significant overlap of the two bands (The *Z*-isomer has a shorter lifetime); (iii) Pseudostilbenes type, which usually contains both electron-donor and electron-acceptor substituents, and nearly degenerate bands (The *Z*-isomer stability is the shortest) [21]. Based on the absorption properties, the photoisomerization step can be induced by a light of the wavelength that falls on either π–π^* (shorter wavelength) or n–π^* (longer wavelength) absorption bands. As a result, the azobenzene can be excited to either $S_2(\pi–\pi^*)$ or $S_1(n–\pi^*)$ states, respectively. If the wavelength is in the overlapping region of absorption for *E*- and *Z*-isomers, both isomerization can simultaneously happen to lead to a photostationary state. [20, 21] Both (*E*)- and (*Z*)-azobenzenes can undergo *E-Z* isomerization involving $S_1 \leftarrow S_0$ and $S_2 \leftarrow S_0$ excitations. However, the sum of the quantum yields for the forward and reverse isomerization steps were found to be constant and also independent of solvent polarity indicating a common excited state for both the photoprocesses [21, 384].

5.1.2. Photoisomerization Mechanism

Unlike stilbene, where only the rotation about the C = C bond is the lone channel at the excited state, the photoisomerization of azobenzene involves various mechanisms (Figure 5.1). Due to the presence of two nitrogen lone pairs apart from the N = N bond, the photoisomerization step has always been debatable. In general, the excitation wavelength influences the isomerization pathways. Typically, the irradiation at π–π* and n–π* bands leads to isomerization through rotation and inversion mechanisms, respectively [385]. The controversy surrounding these two possible mechanisms has been subjected to a lot of experimental and theoretical studies in obtaining the proof. The femtosecond transient absorption measurements revealed two different decay components, namely an ultrafast <200 fs one that can be attributed to motion along the torsional coordinate, and a 2.5-ps lifetime component due to motion along the inversion coordinate under the excitation of the S_2(π–π*) state and for the S_1(n–π*) state, respectively [386, 387]. Through a series of time-resolved absorption and fluorescence spectroscopic and Raman experiments, Fujino et al. interpreted that the entire S_2(π–π*) state population relaxed on a sub-100 fs time scale, and the isomerization mainly happened in the S_1(n–π*) state in ps timescale. Indeed, the Raman spectroscopic studies revealed no substantial change in the frequency of the N = N stretch during the isomerization, which is expected upon double bond breaking along the torsional pathway [388-390]. Evidence through femtosecond time-resolved photoelectron spectroscopy along with *ab initio* molecular dynamics on the excited state dynamics of azobenzene revealed the presence of two near degenerate π–π* excited states (S_2 and $S_{3,4}$). Apart from confirming the inversion mechanism, the wavelength-dependent quantum yield was also explained based on the involvement of the degenerate excited states [391].

Based on the resonance Raman intensity analysis and modeling, inversion and rotation mechanisms have been predicted for the forward (*E* to *Z*) and reverse (*Z* to *E*) photoisomerization channels, respectively [392]. A concerted inversion channel has been proposed based on the fluorescence anisotropy measurements that rationalized both the quantum yields and ultrafast observations, which was further supported by theory as well [393]. Through computations at CASPT2 level, Taketsugu demonstrated that the N = N stretching mode is insensitive to the mechanism [394, 395]. This opened up once again the possibility of a torsional channel. Indeed, a jet-cooled molecular beam and multiphoton ionization techniques in combination with TDDFT level of theory reported high-resolution spectra of S_1(n–π*) and S_2(π–π*). Furthermore, they predicted a barrier in S_1(n–π*) excitation channel with change in the hybridization at nitrogen that supported yet another mechanism inversion-assisted torsional pathway [396]. In a recent work, sub20 femtosecond resolution transient absorption spectroscopy and computations have been utilized to study the ultrafast photodynamics of azobenzene. Not only the $S_2(\pi\text{–}\pi^*)$ lifetime has been estimated to be 50 fs, but it was also correlated with the buildup of $S_1(n\text{–}\pi^*)$, which was responsible for the observed Kasha's rule violation (lower photoisomerization quantum yield for $S_2(\pi\text{–}\pi^*)$ excitation than $S_1(n\text{–}\pi^*)$ excitation) in UV excitation through CNN bending. Through this, the reasons for the lowering of quantum yields at shorter wavelengths, and the ps time scale intramolecular vibrational redistribution from bending into the torsion in attaining *E*-isomer has been confirmed [397]. Apart from these experimental reports, many independent theoretical studies have also been carried out in support of one or the other channels [395, 397-402].

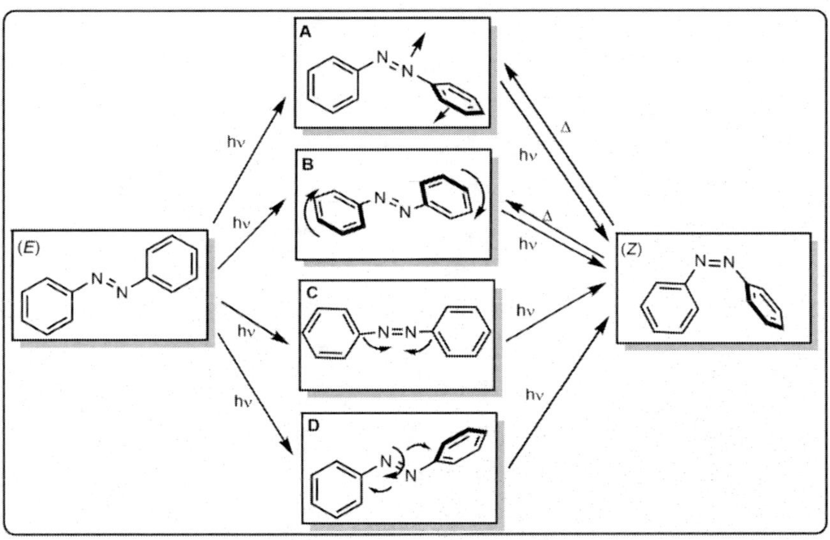

Figure 5.1. Possible mechanisms involved in the photoisomerization and reverse thermal isomerization steps: (A) Inversion [392]; (B) Rotation [392]; (C) Concerted inversion [393] and (D) Inversion assisted rotation [393]. Reproduced with permission from ref. 413; Copyright 2014 American Chemical Society.

5.1.3. Thermal Reverse Isomerization

Various mechanisms have been proposed for the thermal reverse isomerization as well. The initial consideration through substituent effects (Hammett plot) was a rupture, followed by a twist around the –N = N- group, through a singlet transition state. However, the observed preexponential factors (1012-1014 sec^{-1}) excluded the involvement of triplet excited state, else the activation energies are too low for the singlet channel [403]. Based on the observed activation energies (21-24 kcal/mol), another mechanism through destabilization of the double bond via dipolar resonance structures and/or solvent involvement has been considered. Due to the absence of a large solvent or substituent effects, these assumptions have also been ruled out. At this juncture, Talaty proposed an alternate mechanism involving linear transition state, with changes in the hybridization at the nitrogen atoms [404]. Essentially the characteristics such as dipolar, large volume change and the possible involvement of singlet or triplet excited state have been expected for the transition state in the rotational mechanism, whereas a linear TS is expected for the inversion mechanism. Conflicting evidence has been proposed for favoring both mechanisms. Based on the kinetic solvent effects, a rotational mechanism was found to be operative in the push-pull type substituents containing azoarenes [405]. In contrast, systems connected to rigid groups (e.g., crown ethers) with restricted molecular motions can undergo isomerization mainly through the inversion mechanism. Enthalpy of solvent transfer and substituent effects also supported the inversion mechanism [406-408]. Surprisingly, Nishimura et al. observed a V-shaped Hammett plot for *para*-monosubstituted azobenzenes that was accountable for the inversion mechanism, which was controlled mainly through a strong resonance stabilization of its coplanar transition state [408]. Joshi et al. on the other hand, observed in the case of 4-aminoazobenzene that inversion was the preferred mechanistic channel in *Z-E* thermal isomerization in a nonpolar medium; whereas, the isomerization path deviated from the inversion to the rotational route in a polar medium [409]. Thus, both substituent and solvent influence the thermal reverse isomerization kinetics.

In summary, the overall mechanistic aspects related to the various isomerization pathways in azobenzene-based on several important experimental and theoretical findings can be represented as in Figure 5.2. This includes both the light-induced and reverse thermal isomerization steps.

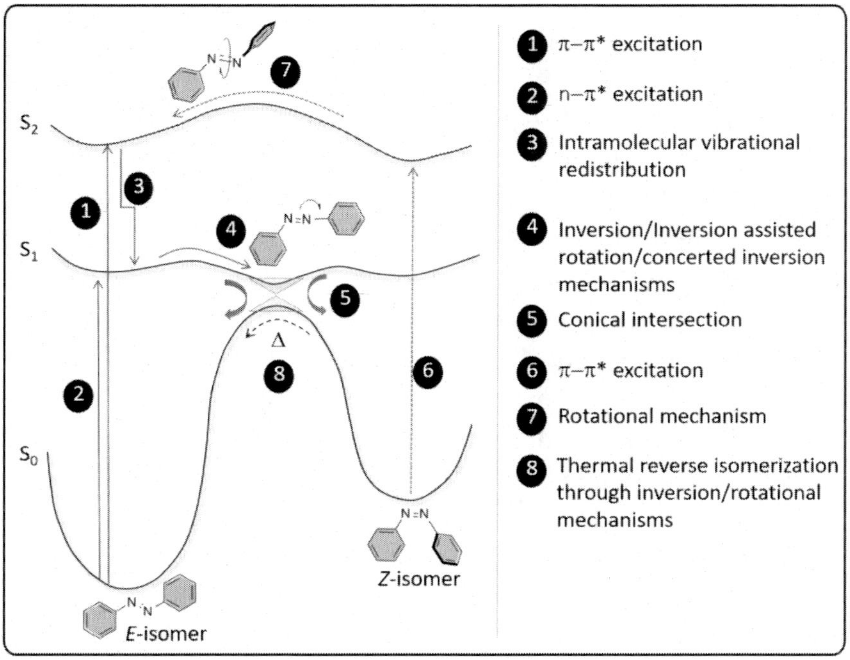

Figure 5.2. Schematic representation of ground and excited state potential energy surfaces illustrating different types of mechanisms involving photochemical and thermal isomerization in azobenzene.

5.2. Photoswitching Behavior in Azoheteroarenes

Unlike the azobenzene, understanding photoswitching behavior of azoheteroarene attracted experimentalists and theoreticians only in recent times. Hence, only limited reports on the systematic investigations with respect to mechanistic aspects, photoswitching efficiencies, photostationary states and also the half-life of *Z*-isomer are available. Majority of those studies are restricted to the monoheterocycle containing azoheteroarenes. Due to the fact that the emergence of five-membered heterocycle derived azoheteroarenes captured widespread attention, the discussion on the photoswitching aspects will be primarily dealt with these systems. However, we generalize the photoswitching behavior, *Z*-isomer stability and also the factors affecting them in a wider context.

5.2.1. Molecular Geometries and Molecular Orbital Analysis

One of the important deviations in comparison to the simple azobenzene is the structural variation in the geometries of the *E*- and *Z*- isomers. Typically, the structures vary from planar (C_{2h} symmetric) *E*-isomer to non-planar (C_2 symmetric) *Z*-isomer in the case of azobenzene. On the other hand, the introduction of heterocyclic moiety not only reduces the symmetry (despite retaining the planarity) of the *E*-isomer but also leads to interesting situations. Particularly when the heterocycle is a five-membered ring, the resulting azoarene can adopt either a twisted or T-shaped geometry. Substitution at the 2-position with respect to the azo group in the five-membered heterocyclic ring can largely influence such geometries (Figure 5.3). Fuchter and coworkers reported the effect of methyl substitution at 2-position in the geometries of *Z*-isomer. In the absence of the methyl group,

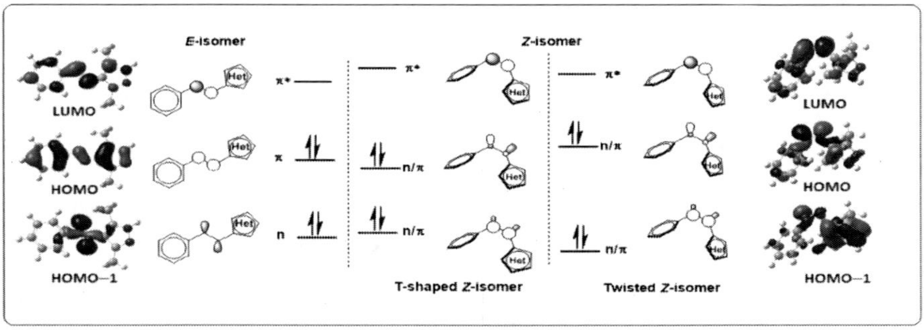

Figure 5.3. Schematic representation of molecular orbitals corresponding to HOMO-1, HOMO, and LUMO of azoheteroarenes. For comparison, the actual MO's of 22 (computed at B3LYP/6-311G(d, p)) rendered at an isovalue of 0.04 are given. (In the case of twisted and T-shaped geometries of the azoheteroarenes, the position of the orbitals are varied in accordance with their trend in energetic ordering with respect to that of *E*-isomer.) Adapted from ref. 150; Copyright 2017 American Chemical Society.

the resulting *Z*-isomer attains a perpendicular T-shaped geometry, in which H-π interaction between the hydrogen atom at 2-position of the ring and the π cloud of the phenyl ring. On introducing a methyl substitution at

2–position, the heterocyclic ring can now orient in two different ways: the H-π interaction can still be possible leading to the T-shaped geometry when the methyl group is oriented outwards; else, it attains a twisted geometry upon the inbound orientation of the methyl group. On the other hand, dimethyl substituted derivatives prefer only a twisted geometry in its *Z*-isomer. The extra stability of the *Z*-isomers in those cases with the possible T-shaped geometries have been attributed to the H-π interaction, which increases the half-life of the *Z*-isomer.

Apart from the geometrical aspects, inspecting the frontier molecular orbitals (FMO) in particular, HOMO-1, HOMO, and LUMO for azoheteroarenes have been thoroughly investigated by various groups [150, 151, 166] (Figure 5.3). Fuchter's group have reported that in the case of *E*-isomer, the HOMO-1, HOMO, and LUMO are corresponding to n, π, and π* orbitals, respectively. On the other hand, the molecular orbitals of *Z*-isomer showed different ordering where HOMO-1, HOMO, and LUMO correspond to π, n, and π* orbitals, respectively [150]. However, the FMOs showed a mixed n and π-character for both HOMO-1 and HOMO in the case of *Z*-isomers. For twisted and T-shaped *Z*-isomer structures, the energies of the FMOs were perturbed and their ordering has been slightly varied relative to their corresponding orbitals in the case of *E*-isomer. Yin et al. predicted significant changes in the energy gaps within HOMO-1 to LUMO and HOMO to LUMO in the cases of nitrogen-containing azoheteroarenes in comparison to azobenzenes [166]. Interestingly, the changes in the gaps are correlating well with the higher stability of *Z*-isomer. Besides that, such energy gaps reflect the trend in the absorption properties. For various *meta*- and *para*-substituted arylazo-2, 5-dimethylpyrazole derivatives, the gap between the HOMO-LUMO (π–π*) and HOMO-1-LUMO (n–π*) reflected the spread in the experimentally observed λ_{max} corresponding to those transitions [151]. It is important to indicate that π–π* transition involves the S_2(π–π*) excited state, the lower HOMO-LUMO energy gaps of the ground state structures are although of the same order as the experimentally observed π–π* transitions, belong to the higher energy transition. Hence, they observed at the shorter wavelength, whereas the n-π* transitions, involving the S_1(n–π*) excited state are found to occur at a longer wavelength.

5.2.2. Photoswitching Behavior in Azoheteroarenes

Generally, photoisomerization phenomena for the azoheteroarenes have been investigated using various spectroscopic techniques. Among them, UV-Vis and NMR are the widely used methods for identifying/characterizing and also the estimation of *E*/*Z* product ratios during the photoisomerization processes. The UV-Vis spectroscopy is based on the different absorbing properties of *E*- and *Z*-isomers in π–π* and n–π* absorptions. In the case of NMR spectroscopy, the upfield shifts in the protons (due to shielding effects of the aryl group on the protons of the adjacent ring) upon isomerization from *E*- to *Z*-isomer will be very helpful. Indeed, the normalized integral ratios of any particular (non-overlapping) proton signal corresponding to *E*- and *Z*-isomers are normally used to follow the kinetics and estimation of PSS. In this section, we have compared the data on photoswitching aspects of various azoheteroarenes. Even though many different types of azoheteroarenes have been reported, only a limited number of reports have actually studied such properties. For understanding the effect of heterocycles, we have taken the Ph-N = N- as a common unit. In this regard, we have compared the absorption and photoswitching properties such as λ_{max}, molar absorptivity, photostationary state (PSS), quantum yield, half-life, and rate constants (Table 5.1). For constructing azoheteroarene based photoswitches, the following three factors need to be considered: stability of *Z*-isomer, tunability and quantitative photoswitching in both forward and reverse isomerization. Based on the available photoswitching and related data on phenylazoheteroarenes compiled in table 5.1, we observed that Fuchter's *N*-methyl phenylazopyrazole **17** with a half-life of 1000 days was the most stable molecule in *Z*-isomeric form. Once the methyl substitutions have been introduced on the pyrazole ring at 3, 5-positions **20**, the half-life tremendously dropped to 10 days. The half-life further reduced to 9.2 min, in case the methyl group at the pyrazole nitrogen is replaced by hydrogen in **22**. The other five-membered heterocycles based azoheteroarenes follow the similar trend with and without the *N*-methyl groups. The same trend was independently demonstrated by König and co-workers utilizing indole-based azoheteroarene derivatives. Moreover, they were also able to tune the half-life by varying the methyl substitution.

Due to the limited number of reports on six-membered azoheteroarenes, comparison and conclusions cannot be obtained at this stage. Similarly, for the photostationary states, there is no direct correlation based on the available structures and forward/reverse photoisomerization conversions. Presumably, the use of monochromatic light source and intensity might play a crucial role in this regard. Various factors influence the stability of the *Z*-isomer and photoswitching aspects of the azoheteroarenes, which are discussed in the proceeding section.

Apart from UV-Vis and NMR spectroscopic methods, Fourier transform infrared (FTIR) spectroscopy has also been used for understanding the changes upon conversion from *E*- to *Z*- isomers and vice-versa. The changes in the geometry and symmetry led to shifts in the IR signals and/or peak splitting in the IR spectra that can be useful in understanding the photoisomerization process. Yu et al. have estimated the *Z*- to *E* thermal reverse isomerization barrier in azobenzene-based on FTIR data, which was well correlated with the theoretically estimated barrier [413]. Quick et al. reported the photoisomerization dynamics and pathways for *E*- to *Z* and *Z*- to *E* using broadband femtosecond transient absorption, fluorescence, stimulated Raman spectroscopy [414]. X-ray diffraction techniques have also been used in understanding the *E-Z* isomerizations [415-417]. Bieske and coworkers have studied the photoisomerization of protonated azobis(2-imidazole) using tandem ion mobility mass spectrometry [418]. Computationally using surface-hopping molecular dynamics at the semi-empirical OM2/MRCI level Pang et al. demonstrated the unidirectional rotary dynamics under a chiral path for *Z-E* photoisomerization step in imidazole-based azoheteroarenes [419].

Table 5.1. Comparison of UV-Vis absorption and photoswitching properties of various azoheteroarenes

S. No.	R - N = N - Ph	UV-Vis absorption properties of *E*-isomer		Photoswitching PSS Data		Half life[a] of *Z*-isomer	Ref.
		$\lambda_{(\pi-\pi*)}$ (nm) / ε (Lmol^{-1}cm^{-1})	$\lambda_{(n-\pi*)}$ (nm) / ε (Lmol^{-1}cm^{-1})	Forward (*E-Z*) /λ (nm)	Reverse (*Z-E*) /λ (nm)		
1	2[c]	385 (22100 ± 1800)	413	84 ± 2 (532)	82 ± 3 (415)	12 h 3 h[b]	150, 50
2	3[c]	394 (28200 ± 1800)	430	85 ± 3 (532)	>98	21 s[b]	150
3	5	363 (19300 ± 1800)	410	77 ± 3 (355)	85 ± 5 (532)	3.7 d[b, c]	150
4	6	345 (16600 ± 1500)	410	85 ± 3 (355)	98 ± 2 (532)	0.25 h[b, c]	150
5	10	318	454	60 (382)	–	–	171
6	11	318	426	58 (365)	15 (405)	–	153
7	13	336 (14800)	–	>95 (365)	–	2.4 d[b] 4.7 h	49, 150
8	14	363 (17000)	450 (1260)	>98 (363)	–	9 h[b]	150, 410, 411
9	15	320 (14600 ± 1700)	425 (778 ± 91)	>98 (355)	97 ± 3 (532)	74 d	150
10	16	341 (16200 ± 2000)	425 (676 ± 81)	>98 (355)	79 ± 3 (532)	11 d	150
11	17	328 (16600 ± 1300)	417 (640 ± 51)	>98 (355)	70 ± 3 (532)	1000 d[a]	50, 150
12	19	362 (25800)	–	98 (365)	45 (455)	6.5 d[b] 22 d	49, 150
13	20	335 (22700 ± 1800)	425 (963 ± 76)	>98 (355)	>98 (532)	10 d	50, 150
14	21	325 (17400 ± 2000)	428 (612 ± 72)	96 ± 2 (355)	87 ± 3 (532)	5.8 d	150
15	22[c]	330 (19369 ± 278)	421 (1199 ± 57)	85 (365)	77[f]	9.2 ± 2.3 min	151
16	25	317	426	59 (365)	12 (405)	–	153
17	27	328	433	81 (365)	14 (460)	36h[b]	153
18	28	310 (18700 ± 2200)	438 (505 ± 59)	90 ± 2 (355)	84 ± 3 (532)	21 d	150
19	29	310 (16700 ± 1500)	442 (373 ± 34)	76 ± 2 (355)	72 ± 3 (532)	39 d	150
20	63 (*N*-H) 63a (*N*-THP)	362 363	– 442	– 89(365)	– 24(460)	– 19h[b]	153
21[c]	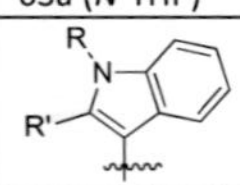	372 (16447) **64a**: R = H, R′ = H 379 (14658) **64b**: R = Me, R′ = H 380 (15964) **64c**: R = H, R′ = Me	–	85[d] (400)	–	6.5 s[b] 2.6 d[b] 1.3 ms[g]	149
22	65	349	428	–	–	1.8 min[b]	153
23	66(*N*-H)	349	430	96 (365)	35 (460)	–	153

S. No.	R - N = N - Ph	UV-Vis absorption properties of *E*-isomer		Photoswitching PSS Data		Half life[a] of *Z*-isomer	Ref.
		$\lambda_{(\pi-\pi^*)}$ (nm) / ε ($Lmol^{-1}cm^{-1}$)	$\lambda_{(n-\pi^*)}$ (nm) / ε ($Lmol^{-1}cm^{-1}$)	Forward (*E-Z*) /λ (nm)	Reverse (*Z-E*) /λ (nm)		
	66a(*N*-Tos)	335	430	95(365)	26(460)		
24	67	325	–	–	–	–	153
25	69	336	432	92 (365)	21 (460)	–	153
26	76(*N*-H) 76a(*N*-Tos)	355 338	433 431	95 (365) 96(365)	33 (460) 27(460)	3h[b]	153
27	77(*N*-H) 77a(*N*-Tos)	387 351	435 435	88 (405) 96(365)	21 (523) 28(460)	–	153
28	79	290 (14454)	–	–	–	41 min[b]	279
29	80	288 (15135)	–	–	–	5 s[b]	279
30	82	285 (12882)	–	–	–	<4 s[b]	279
31	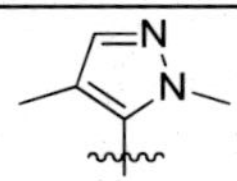	340 (18600 ± 2100)	435 (854 ± 95)	94 ± 2 (355)	86 ± 3 (532)	12 h[b]	150
32	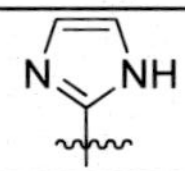	362 (18000)	450 (960)	–	–	–	410
33	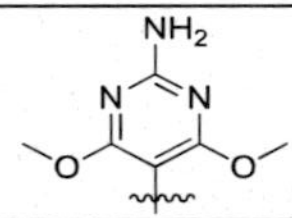	352 (24700[d])	440 (2630[d])	20 (370)	59 (465)	223.6 h	412

[a]NMR; [b]UV-Vis; [c]sensitive to water present; [d]DMSO; [e]MeCN; [f]white light; [g]nanosecond laser flash photolysis. Values adapted with permission from ref. 50; Copyright 2016 American Chemical Society; Values adapted with permission from ref. 82 Copyright 2016 American Chemical Society; Values adapted with permission from ref. 279 Copyright 2016 American Chemical Society. (THP = Tetrahydropyran; Tos = Tosyl)

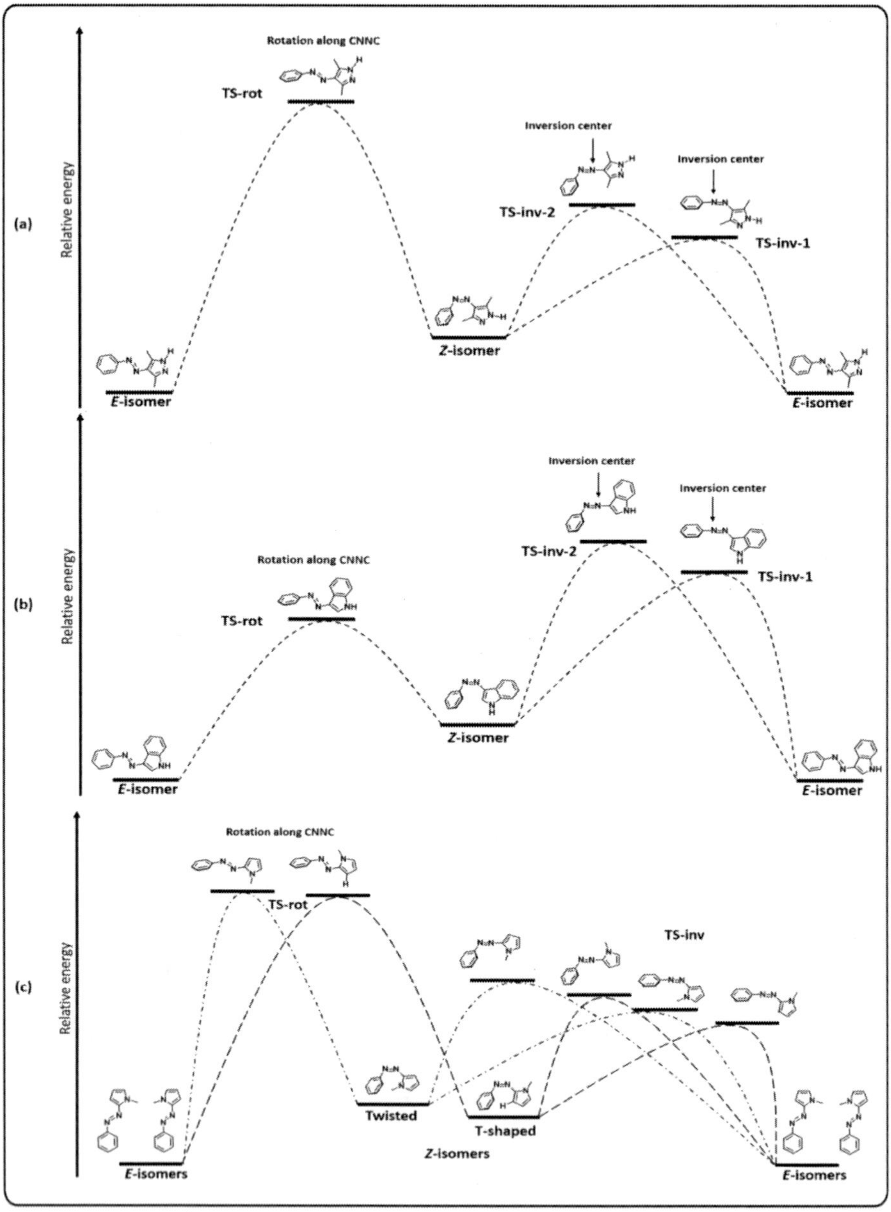

Figure 5.4. Inversion and rotational barriers in the *Z*-*E* isomerization pathways of (a) phenylazo-2, 5-dimethylpyrazole [151], (b) phenylazo-*N*-methylindole [149] and (c) phenylazo-*N*-methylprazole [150]. Reproduced with permission from references 149, 150 and 151; Copyright 2017, 2018 American Chemical Society.

5.2.3. Stability of Z-Isomer and the Thermal Barriers for the Reverse Isomerization

Synthesis and characterization of azoarenes based photoswitches having very high stability and long half-life in *Z*-isomer still remains challenging. Introduction of heterocycles (as azoheteroarenes) found to be one of the viable methods in increasing the half-life of *Z*-isomer tremendously. For instance, the *Z*-isomer of azoheteroarenes with *N*-methylated imidazole [49] was reported to have a half-life of 528 h, whereas *N*-methyl pyrazole [50] was found to have a half-life up to 1000 days. The stability of the *Z*-isomer completely depends on the thermal barrier connecting the *Z*- and *E*-isomers in the PES. As indicated before two different mechanisms, namely inversion and rotation are possible for azoheteroarenes (Figure 5.4). Both the isomerization mechanisms happen through the changes in either the dihedral or the bending modes associated with the azo group. The rotational mechanism involves the rotation along the central CNNC dihedral angle, whereas, the inversion happens at the nitrogen atom of the azo group through bending of CNN or NNC angles. Since many of the reported azoheteroarenes are unsymmetrical, two different inversions and one-rotation channels are possible.

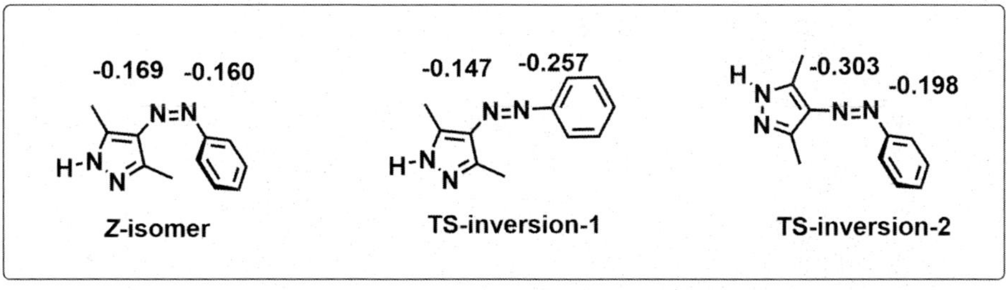

Figure 5.5. Charges at the nitrogen atoms in the inversion transition states relative to the Z-isomer (calculated at B3LYP/6-311G(d, p)) [151].

Depending on the groups and substituents, one of the inversion transition states is often found to be low-lying. In the case of arylazopyrazoles, phenyl ring connected nitrogen inversion transition state was found to have a lower energy barrier compared to pyrazole connected nitrogen inversion transition state. In the inversion mechanism, accumulation of more negative charge at the inverting center in the transition states relative to the corresponding *Z*-isomer was observed (Figure 5.5). Zhang and co-workers have demonstrated that the electron donating groups (EDG) provides an advantage for the inversion mechanism during the thermal reverse isomerization in azoheteroarenes [166]. Based on the reported mechanistic aspects of various azoheteroarenes, inversion mechanism was found to be the dominant channel. This inversion mechanism was operative for systems such as azopyrazoles, azopyrrole, and azoimidazole etc. However, arylazoindoles exhibited rotation as a minimum energy mechanism, whereas the inversion barrier was found to be 10.4 kcal/mol higher in energy [149]. In yet another case, arylazoindazoles also exhibit rotational channel as the minimum energy route [366]. The energy barrier for the *Z-E* isomerization and their corresponding HOMO to LUMO, HOMO-1 to LUMO energy gaps provide useful insights into the stability of *Z*-isomer.

5.2.4. Factors Affecting the Z-Isomer Stability and Photoswitching Behavior

It is well established that azobenzene and its derivatives undergo a thermal *Z-E* isomerization on the ground state potential energy surface. Controlling the photoswitching and tuning of the *Z*-isomer stability are very relevant to obtain various applications. Thus, apart from understanding different pathways in the thermal reverse isomerization channels, transition states, to envisage various factors influencing the barriers and/or the stability of *Z*-isomer is critical. Experimental and theoretical reports are available in demonstrating the influence of steric and electronic factors, role of solvents, and concentration dependency on these aspects. Apart from that, the effect of hydrogen bonding is also very important in this regard. Herein, various factors influencing such aspects are discussed.

(a) *Steric factor:* In general, the *ortho* substitution either at the aryl ring or the heterocyclic ring can destabilize the *Z*-isomer of azoheteroarenes. Fuchter and coworkers reported that steric factor in azoheteorarenes due to *ortho*-methyl group(s) was responsible for the *Z*-isomers in attaining a twisted geometry [150]. The twisted geometry of the *Z*-isomer is found to be more destabilized, as a result, they can readily undergo thermal reverse isomerization with lower half-life. On the other hand, the systems with monomethyl group can adopt either a twisted (destabilizing) or T-shaped (stabilizing) geometry. Overall, the weighted average half-life will be influenced by the difference in energy between the two possible *Z*-isomeric geometries and also the barriers relative to them (Figure 5.4c). Devi et al. have observed all the *E*-isomers of the bis(*ortho*)-aryl substituted azo-2, 5-dimethylpyrazole derivatives, the changes in the π–π* spectral features showed blue shifts relative to the unsubstituted one [151]. This was explained on the basis of the steric factor, which affects the planarity and conjugation. However, they observed redshifts in the case of the mono(*ortho*)-aryl substituted systems. In the same work, they have also reported a negative slope in the Taft plot corresponding to the thermal reverse isomerization kinetics of mono(*ortho*)-aryl substituted azo-2, 5-dimethylpyrazole derivatives. The steric release in attaining transition state relative to the Z-isomer was accounted for this observation. Through theoretical calculations on five-membered azoheteroarenes, Zhang and coworkers have predicted that the steric factors influence in two ways: (1) the separation between the π–π* and n–π* bands increase; (2) the inversion barrier corresponding to thermal reverse isomerization lowers in energy [166]. Without the involvement of solvent model, using molecular dynamics Zhou et al. predicted a difference in the mechanisms for N-H and N-methylated arylazoimidazoles [420].

(b) *Electronic effects:* Traditionally the electronic effects have been understood from Hammett plots. In order to understand the electronic effects in the *Z*-isomer stability, reverse isomer kinetics have been done using substituted arylazoheteroarenes. Devi et al. have performed the kinetics experiments using arylazopyrazoles [151]. Two sets of experiments for *meta*- and *para*- substituted ones at two different concentrations (μM and mM) confirmed the presence of substantial electronic effects. However, the slopes were found to be completely opposite indicating the presence of complex interplay of concentration dependency and various other factors such as hydrogen bonding, tautomerism etc. (vide infra). However, in the same work, the computations predicted a very good correlation in the linear free energy relationship with respect to electronic effects for both *meta* and *para* substituted azopyrazoles. In another work on the *para*-substituted 5-phenylazopyrimidines, Čechová et al. have demonstrated the effect of substituents on the photoisomerization and the *Z-E* thermal reverse isomerization [395]. They have computationally predicted that the inversion barrier in which the nitrogen inversion center connected to the *para*-substituted phenyl showed a substantial effect than the other nitrogen inversion center that was connected to the pyrimidine. Interestingly, for the electron donating and the electron withdrawing substitutions at *para* position, the lower energy inversion mechanistic channels are at the nitrogen center attached to the pyrimidine and the nitrogen center attached to the *para* substituted aryl ring, respectively.

(c) *Hydrogen bonding, tautomerism, and solvent effects:* In the case of azoheteroarenes based photoswitchable systems, concentration dependency in the photoisomerization, as well as thermal stability of the *Z*-isomer, have been reported. Apart from that, the presence of water also influenced the photoisomerization and strongly destabilizes the *Z*-isomer by increasing the reverse isomerization kinetics [50]. Based on the photo- and thermal isomerization of arylazoimidazoles, Otsuki et al. have estimated higher quantum yield and fast thermal isomerization rates than in azobenzene. They attributed the fast thermal isomerization to the self-catalysis and protic solvent-assisted tautomerism [410]. Devi et al. have estimated differences in thermal reverse isomerization kinetics at μM and mM concentrations [151]. All these results indicate the presence of hydrogen bonding and its influence. Particularly when the molecules having free NH (such as pyrazole, imidazole etc.) can self-dimerize by donating the hydrogen either to the azo group or the acceptor nitrogen of the heterocyclic ring. As a result of the accumulation of the charge in the heterocyclic ring due to the hydrogen bonding, the push-pull type effect persuades that results in the fast reverse thermal isomerization (Figure 5.6a). At a lower concentration, these channels may subside, however, the presence of protic solvents such as water can assist in tautomerism, which dominates the reverse thermal isomerization step. This can be clearly understood from the work by Devi et al. on the solvent effects in the reverse thermal isomerization kinetics in phenylazopyrazole and *N*-methyl phenylazopyrazole [151]. The former showed substantial solvent effects due to solvent-assisted tautomerism, whereas even at a very high concentration, the later did not show much variation in the rate constants (Figure 5.6b). Similarly, Velasco and coworkers predicted that tautomerism played an important role in the fast switching of ionic (methylated) pyridinium based azo compounds [421].

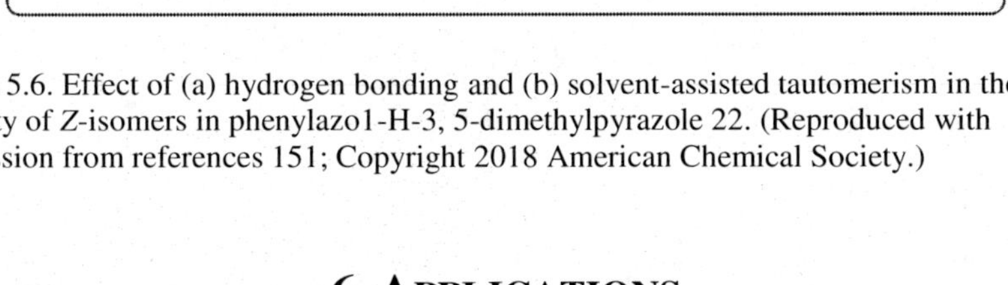

Figure 5.6. Effect of (a) hydrogen bonding and (b) solvent-assisted tautomerism in the stability of *Z*-isomers in phenylazo1-H-3, 5-dimethylpyrazole 22. (Reproduced with permission from references 151; Copyright 2018 American Chemical Society.)

6. APPLICATIONS

In the historical perspectives, the azoheteroarenes have been mainly utilized as dyes. However, they turned out to be potential candidates for applications in different fields such as medicinal chemistry, supramolecular chemistry, catalysis etc. Many such applications are still under evaluation.

Ever since the introduction of the light-induced photoisomerization, relevant applications such as photoswitches, and data storage have also emerged. We mainly discuss some of the most interesting applications related to the azoheteroarenes in the past decades. More emphasis has been given to the recent literature.

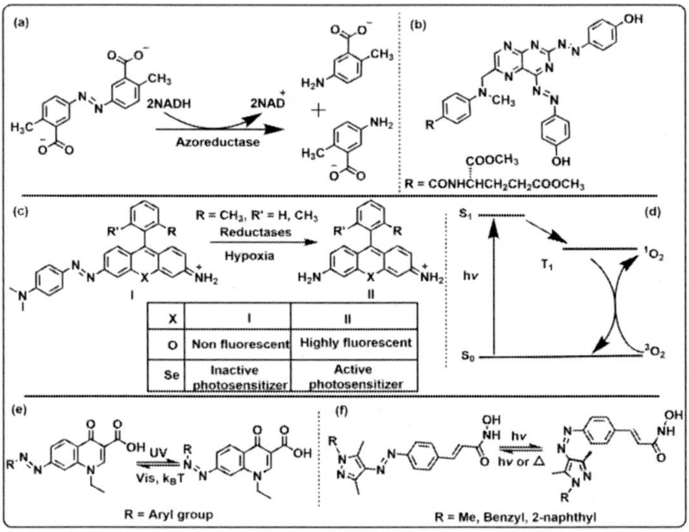

Figure 6.1. Representative examples for the applications of azoheteroarenes in medicinal chemistry. (a) Involvement of azoreductase enzyme in the cleavage of azoarenes into amines; (b) Methotrexate based azo derivative as a colon cancer drug; (c) and (d) Smart fluorescent and activatable photosensitizer (Reproduced with permission from ref. 8; Copyright 2017 American Chemical Society); (e) Azofunctionalized antibiotics (Reproduced with permission from ref. 424; Copyright 2013 Springer Nature); (f) Photoswitchable inhibitors targeting bacterial HDAC. (Reproduced with permission from ref. 425; Copyright 2017 American Chemical Society.)

6.1. Medicinal Chemistry

In the formative years, various elegant synthetic methods have been adopted in making various azoarene connected heterocycles and evaluated for their medicinal aspects in particularly the anti-infectives therapeutic areas [196]. However, the pharmaceutical effect for many such azoheteroarenes was found to be due to the reduction, followed by cleavage of azo group [422]. Azoreductase, a flavoenzyme in a range of prokaryotes and eukaryotes was found to be responsible for that. The mechanism of reduction requires tautomerization, and so many azoarenes with hydroxy substitution either at 2- or 4- position with respect to the azo groups have been utilized in this regard (Figure 6.1a). Using this principle, many prodrugs have also been generated for various therapeutic areas. For instance, methotrexate (in Figure 6.1b), a colon cancer drug has been modified into 4-hydroxyphenyl azo derivative, that exhibited 60-70% release after 24h [423]. The identical hypothesis has also been used to photosensitize oxygen (to generate singlet oxygen) in selectively ablating the tumor cells under hypoxia, a common condition in solid tumors. Urano and coworkers in their remarkable work come up with a selenium or oxygen-based dye that has an azo functionalization (in Figure 6.1c and d). Interestingly, they demonstrated smart fluorescent and activatable photosensitizer properties for oxygen and selenium-containing species, respectively [8]. Feringa and coworkers have utilized the light-driven properties towards temporal control of the availability of azoquinolones in obtaining the pharmacodynamic aspects [424]. In the demonstration of the proof-of-concept, *Z-E* isomerization in azofunctionalized antibiotics has been fully utilized in controlling the bacterial growth by light (Figure 6.1e). The inactive *Z*-isomer of the drug upon converted back to *E*-form was activated. Following this, Fuchter's group have developed drug candidates using azobenzene or azopyrazole as photoswitchable inhibitors targeting bacterial HDAC [425]. They observed a difference in the activity between *E*/*Z*-isomers up to a maximum of 10-fold (Figure 6.1f). Apart from the simple azoheteroarenes, several metal complexes of them have also been utilized for various aspects in medicinal chemistry such as photosensitizers in photodynamic therapy (PDT) [426], antiproliferative activity towards cancer cells and distinctive knotting of DNA [119, 427, 428] and potential glutathione (GSH)-responsive photoactivated chemotherapy (PACT) agent [121, 429]etc. Except for phototrexate [429], in all other cases, azopyridine based ligands have been utilized as metal complexes, however, the biological activities were evaluated at its native state. In a photopharmocological application, a sulfonylurea-based azoheteroarene has been utilized to control over pancreatic beta cell function and insulin release optically. The salient feature of this application is the use of yellow light activation due to the availability of push-pull type system [196].

6.2. Supramolecular Chemistry

Ravoo and coworkers have extensively explored the light responsive hydrogels, micelles and vesicles, surfaces and drug delivery vehicles based on the host-guest complexes of cyclodextrin and cucurbit[8]uril with azoheteroarenes [152, 430-432]. In the proof-of-principle, they have demonstrated the light-induced aggregation and dispersion behavior in cyclodextrins. They have also extended the study using NIR light induced photoswitching by utilizing lanthanide doped nanoparticles and light responsive self-assembly of colloidal molecules [188, 433]. In yet another application related to the nanotechnology, Adam et al. have utilized the azopyrazoles to open and close the DNA hairpin by light [434]. Azopyridines as cocrystals with various carboxylic acids showed interesting supramolecular interactions and physical properties [133, 435]. Similarly, Giese and coworkers have investigated the supramolecular aggregates with liquid crystalline properties using azopyridines [436]. Also, photoresponsive liquid crystals with sematic phase to isotropic mesophase changes have been reported for azopyridines with halogen bonding [437]. Furthermore, the azopyridine based thermal and photoresponsive self-healable soft crystals have also been reported that showed bending and twisting properties [126, 438]. Bushuyev et al. have reported azopyridine based cocrystals with haloazoarenes that showed photo-mechanical properties and crystal-to-crystal isomerization [132]. In the light responding molecular ground, the phototweezer molecule developed by Shinkai et al. using a bis(crown ether) connected azoarene that showed reversible light responsive binding to metal ions such as Rb^+, Cs^+, K^+ etc [439]. This work is still considered to be one of the important contributions of azobenzene in supramolecular chemistry. Apart from these, several metal- and covalent organic frameworks (MOF and COF) have utilized the azoheteroarenes as integral part and linker etc.

Such azoheteroarenes (mainly azopyridine) offered a variety of novel architectures benefits in obtaining pillar-layer [144], 3D-pillar [440], hybrid microphorosity [441], π–π interaction [442] etc., which in turn, provided high selectivity in absorption [441], photoresponsive emission [443], and catalysis [444].

6.3. Catalysis

Light-controlled molecular function provides several applications including the development of smart devices and materials. In a remarkable work, Hecht and coworkers devised a strategy to control the catalysis power by photoswitching of azoarenes (Figure 6.2a). Essentially, they have connected a piperidine base, which was shielded under *E*-isomeric state of the azobenzene, however, it can be accessible upon *E-Z* isomerization. Using light modulated on and off state of the base, they have demonstrated the tuning of the rates of a base catalyzed Henry reaction [445, 446]. Freixa and coworkers have come up with ruthenium-based azoheteroarene complexes in the generation of hydrogen under catalytic conditions using ammonia-borane adduct as a hydrogen source (Figure 6.2b). Despite the inhibition of coordination upon *E-Z* isomerization in complexes with phenylazopyridine as a ligand, they demonstrated light-controlled hydrogen release using bipyridyl connected azobenzenes, which also showed photoswitching. Azopyridine-based ruthenium(II) catalysts have been used for light-controlled hydrogen generation [124]. (Figure 6.2c).

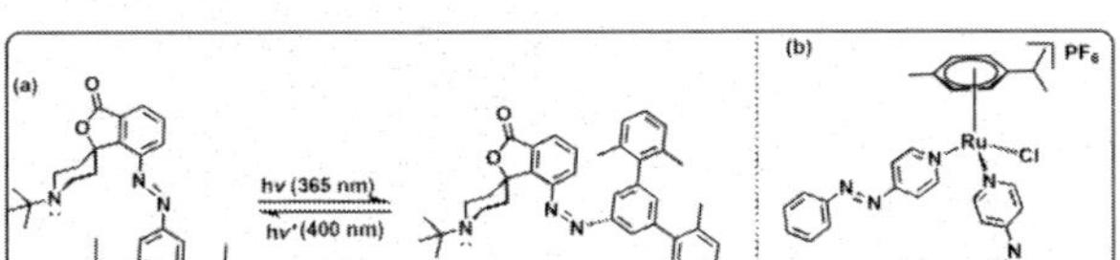

Figure 6.2. Representative examples of azoheteroarenes in catalysis applications. (a) Photoswitchable base catalyst (Reproduced with permission from ref. 446; Copyright 2009 American Chemical Society) and (b) Photoswitchable ruthenium-based dehydrogenative catalyst.

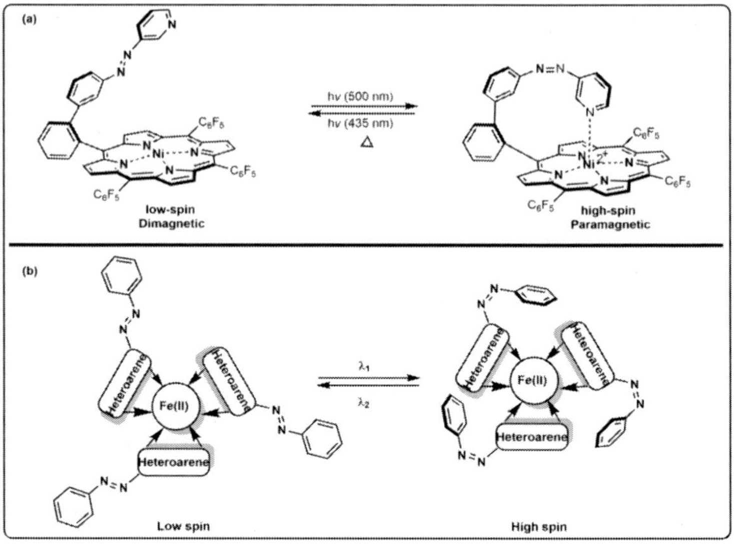

Figure 6.3. Representative approaches to magnetic spin state changes using azoheteroarenes: (a) coordination-induced spin-state switch (LD-CISSS) examples of azoheteroarenes in magnetism applications: Reproduced with permission from ref. 147; Copyright 2015 American Chemical Society. (b) Light-driven ligand-induced spin change (LD-LISC) [452].

6.4. Spin Crossover and Magnetic State Switching

One of the most widely attempted applications in utilizing azoarenes is the data storage. The possibility of molecular logic gates due to light-induced reversible isomerization and associated properties change provided such options. Different options such as light and pH have been used as input and the changes in the magnetic properties like diamagnetic to paramagnetic changes have been utilized for reading out [161]. In a remarkable design, Herges and coworkers have developed a "record player" (nickel porphyrin tethered with phenylazopyridine) as a molecular machine that can exhibit light-induced magnetic properties change in the solution form at room temperature [145, 161, 162, 447-449](Figure 6.3a). The coordination-induced spin-state switch (LD-CISSS) has been fully utilized in reversibly converting the molecule between diamagnetic and paramagnetic states. Furthermore, they have demonstrated the utility of such molecule as a switchable MRI contrast agents [450]. Utilizing many azoheteroarenes, Herges group has improved the switching efficiency and magnetic contrast. Apart from this light-induced changes in the coordination number, approaches based on light-induced excited spin state trapping (LIESST) [451] and light-driven ligand-induced spin change (LD-LISC) [452] have also been utilized in getting the magnetic properties to change. In this regard, Herges and co-workers have utilized azopyridine/azobipyridine based ligands with iron(II) [178](Figure 6.3b). The light-induced tuning in the conjugation of azoheteroarene ligands and the resulting changes in the properties of ligand strength are the underlying principles in such cases.

6.5. Dyes, Photochromism, Sensors, and Sensitizers

Despite the continual use of azoheteroarene in the traditional dye applications, in recent decades, they have shown extended use in NIR dyes (Figure 6.4a). These dyes need long wavelength absorbing azo chromophores. The push-pull type molecules do absorb in such regions, which can relax at a faster rate, and so they have lower *Z*-isomer population. Few azoheteroarenes have exhibited such NIR absorbing properties [453]. Li et al. have synthesized pyrroles with mono and bis azoarenes and also their BF_2 adduct and reported their near infrared absorbing dye properties [454]. Several fused azoheteroarenes such as benzothiazole, benzoselenozole indole, and quinoline have exhibited strong near-infrared absorbance in the range 700-800 nm due to the cationic delocalization, which makes them as very useful azo dyes [455]. Such molecules can undergo photoisomerization in vivo and can be potential photo-pharmaceuticals [456]. Few BF_2 coordinated azo-switches have also shown visible light and NIR activation that makes such molecules useful in biological samples [457, 458]. Many other dye molecules based on azoheteroarenes have been synthesized and reported [459-461]. Apart from these, the azoheteroarene based molecules exhibited solid state photoswitching. Tris-arylazopyrazole functionalization with trimesic acid led to the generation of several C_3 symmetric molecules that exhibited photoswitching in a solid state as well as in solution phase (Figure 6.4b). Interestingly, the molecules showed excellent *Z*-isomer stability and variation in the molecular crystallinity. The polycrystallinity changed to amorphous sample upon irradiation. Remarkably, the contrasting color changes associated with those molecules provided applications in the field of reversible image printing and erasing [462] (Figure 6.4c). Recently, Sinha and coworkers have reported the photochromism of metal-coordinated azoimidazole based compounds exhibiting color changes upon irradiation [463]. Certain polymers with azopyridine moieties are known to influence, complexation-sensors and photo-responsive comonomeric azopyridine (APMA) ligand possessing complexation sensitivity upon addition of CoS to the solution that can inhibit *E*/*Z* isomerization [464]. Horvath et al. have utilized isatin pentafluorophenylhydrazones in anion sensing. The proton abstraction by the anion enables the *E-Z* isomerization in this system, which otherwise a fast switching molecule due to hydrogen bonding [465]. Khurana and coworkers have developed an ampyrone based azo dye, a pH-responsive and also a fluorescent probe for the detection of Al^{3+} in semi-aqueous medium [466].

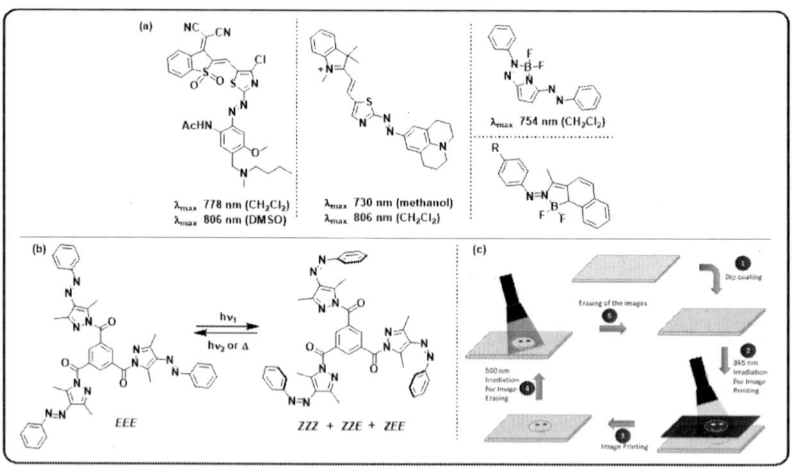

Figure 6.4. Representative examples of azoheteroarenes (a) as NIR based dyes (Reproduced with permission from ref 456 Copyright 2015 American Chemical Society), and (b) as photochromic material for reversible and rewritable image printing and erasing applications (Reproduced with permission from ref 462 Copyright Wiley-VCH Verlag GMBH & Co.).

6.6. Miscellaneous Applications

Apart from the above-mentioned applications, azoheteroarenes have been utilized in a variety of other applications. Particularly the metal complexes of azoheteroarenene have been screened for their anticancer [119], DNA knotting [120, 427], photoactivated chemotherapy [429], reversible electro-polymerization [467], antibacterial [468], cytotoxic [123], and light-controlled hydrogen generation properties [124] and energy storage properties [204-208] etc. In yet another application, the salts of 5, 5'-azotetrazoles gained considerable interest as potential explosives [343, 344], owing to the higher nitrogen content. In addition, they have found a number of practical applications like gas generators for airbags and also for fire extinguishing systems [345-348].

Conclusion and Perspectives

Through this chapter, a brief description of the status of azoheteroarenes has been discussed. Despite a very long history of azoheteroarenes, their photochemistry is still in the emerging phase. Particularly, in the past two decades the amount of publications on these topics reveals that the interest in azoheteroarene has not subsided at all. Over these periods, a wide range of synthetic strategies has also been developed in obtaining various azoheteroarenes. As discussed earlier, the azoheteroarenes provide a lot of benefits in the design and development of new photoswitchable molecules with different functionalities. The presence of heteroatoms offers the possibilities of hydrogen bonding and metal coordination that in turn provides a lot of prospective applications. Among the several reported azoheteroarenes, the five-membered ring containing systems, in particular, azopyrazole derivatives attained a widespread attention. The ease of synthesis, extremely stable *Z*-isomer, and quantitative photoswitching in both forward and reverse isomerization steps are their salient features. Various factors such as steric, and electronic effects through substituents, and hydrogen bonding along with solvent-assisted tautomerism strongly influence the stability of the *Z*-isomer of the azopyrazoles. Both hydrogen bonding and solvent effects need to be carefully considered for improving the photoswitching. In contrast, the corresponding six-membered azoheteroarenes are relatively less explored. Considering the perspective applications in a wide range of fields, the exploration on several azoheteroarenes towards understanding the photoswitching in the fundamental point of view, further improvement in the design are currently the most important challenges ahead. Moreover, the incorporation of heterocycle not only improves the stability and tunability of the *Z*-isomers in azoheteroarenes but also provides ample opportunities in creating many potential applications, where azobenzenes play an integral part. Due to the fact that the majority of the reports on azoheteroarenes did not utilize the photoswitching phenomena, there will be a tremendous scope for further improvement in various fields, where photoswitches can provide optical control. With the possibility of widespread applications, one can expect a potential growth in the chemistry of azoheteroarenes.

ACKNOWLEDGMENTS

We are grateful to the Science and Engineering Research Board (SERB), New Delhi for financial support (EMR/2014/000780). We are also thankful to IISER Mohali for the start-up grant, and necessary infrastructure. M.S. thank CSIR for JRF and SRF Research fellowships (09/947(0065)/2015-EMR-I). A.K.G., S.G. and D.G. thank IISER Mohali for the research fellowships.

The authors S.G., D.G., A.K.G., and M.S. have equally contributed to the chapter.

REFERENCES

[1] Bafana, A.; Devi, S. S.; Chakrabarti, T. *Environ. Rev.* 2011, *19*, 350-370.

[2] Welham, R. D. *J. Soc. Dyers Colour.* 1963, *79*, 98-105.

[3] Roger Finch, D. G., Richard Whitley, S. Ragnar Norrby *Antibiotic and Chemotherapy-9th Edition*; Saunders Elsevier, 2011.

[4] Lesch, J. E. *The First Miracle Drugs: How the Sulfa Drugs Transformed Medicine*; Oxford University Press, 2006.

[5] Yousaf, A.; Shafida Abd, H.; Umer, R. *Mini. Rev. Med. Chem.* 2018, *18*, 1548-1558.

[6] Dembitsky, V. M.; Gloriozova, T. A.; Poroikov, V. V. *Nat. Prod. Bioprospect.* 2017, *7*, 151-169.

[7] Walker, R. *Food Chem. Toxicol.* 1970, *8*, 659-676.

[8] Piao, W.; Hanaoka, K.; Fujisawa, T.; Takeuchi, S.; Komatsu, T.; Ueno, T.; Terai, T.; Tahara, T.; Nagano, T.; Urano, Y. *J. Am. Chem. Soc.* 2017, *139*, 13713-13719.

[9] Piao, W.; Tsuda, S.; Tanaka, Y.; Maeda, S.; Liu, F.; Takahashi, S.; Kushida, Y.; Komatsu, T.; Ueno, T.; Terai, T.; Nakazawa, T.; Uchiyama, M.; Morokuma, K.; Nagano, T.; Hanaoka, K. *Angew. Chem. Int. Ed.* 2013, *52*, 13028-13032.

[10] Zeng, L.; Kuang, S.; Li, G.; Jin, C.; Ji, L.; Chao, H. *Chem. Commun.* 2017, *53*, 1977-1980.

[11] Thiele, J.; Uhlfelder, E. *Justus Liebigs Ann. Chem.* 1898, *303*, 93-107.

[12] Yilmaz, M. D.; Oktem, H. A. *Anal. Chem.* 2018, *90*, 4221-4225.

[13] Hickman, K. C. D.; Linstead, R. P. *J. Chem. Soc., Trans.* 1922, *121*, 2502-2506.

[14] Diehl, H.; Ellingbee, J. *Anal. Chem.* 1960, *32*, 1120-1123.

[15] Anderson, R. G.; Nickless, G. *Analyst* 1967, *92*, 207-238.

[16] Peters, C. A.; Redmon, B. C. *J. Chem. Educ.* 1940, *17*, 525.

[17] Hartley, G. S. *Nature.* 1937, *140*, 281.

[18] Fischer, E.; Frankel, M.; Wolovsky, R. *J. Chem. Phys.* 1955, *23*, 1367-1367.

[19] Hamon, F.; Djedaini-Pilard, F.; Barbot, F.; Len, C. *Tetrahedron* 2009, *65*, 10105-10123.

[20] Merino, E.; Ribagorda, M. *Beilstein J. Org. Chem.* 2012, *8*, 1071-1090.

[21] Bandara, H. M. D.; Burdette, S. C. *Chem. Soc. Rev.* 2012, *41*, 1809-1825.

[22] Merino, E. *Chem. Soc. Rev.* 2011, *40*, 3835-3853.

[23] Harris, J. D.; Moran, M. J.; Aprahamian, I. *Proc. Natl. Acad. Sci.* 2018, *115*, 9414.

[24] Erbas-Cakmak, S.; Leigh, D. A.; McTernan, C. T.; Nussbaumer, A. L. *Chem. Rev.* 2015, *115*, 10081-10206.

[25] Baroncini, M.; Bergamini, G. *Discovering the Future of Molecular Sciences*, Wiley-VCH Verlag GmbH & Co., 2014.

[26] Zhao, Y. Tomiki Ikeda *Smart Light-Responsive Materials: Azobenzene-Containing Polymers and Liquid Crystals*, John Wiley & Sons, Inc. 2009.

[27] Xia, X.; Yu, H.; Wang, L.; ul-Abdin, Z. *RSC Adv.* 2016, *6*, 105296-105316.

[28] García-Amorós, J.; Velasco, D. *Beilstein J. Org. Chem.* 2012, *8*, 1003-1017.

[29] Jeong, S. P.; Renna, L. A.; Boyle, C. J.; Kwak, H. S.; Harder, E.; Damm, W.; Venkataraman, D. *Sci. Rep.* 2017, *7*, 17773.

[30] Russew, M.-M.; Hecht, S. *Adv. Mater.* 2010, *22*, 3348-3360.

[31] Beharry, A. A.; Woolley, G. A. *Chem. Soc. Rev.* 2011, *40*, 4422-4437.

[32] Qin, C.-G.; Lu, C.-X.; Ouyang, G.-W.; Qin, K.; Zhang, F.; Shi, H.-T.; Wang, X.-H. *Chin. J. Anal. Chem.* 2015, *43*, 433-443.

[33] Rasmussen, P. H.; Ramanujam, P. S.; Hvilsted, S.; Berg, R. H. *J. Am. Chem. Soc.* 1999, *121*, 4738-4743.

[34] Wegener, M.; Hansen, M. J.; Driessen, A. J. M.; Szymanski, W.; Feringa, B. L. *J. Am. Chem. Soc.* 2017, *139*, 17979-17986.

[35] Nakasone, Y.; Ooi, H.; Kamiya, Y.; Asanuma, H.; Terazima, M. *J. Am. Chem. Soc.* 2016, *138*, 9001-9004.

[36] Zhang, F.; Zarrine, A.;-Afsar, A. Z.;-Wahid, M. S. A.; Prosse, R. S.; Davidson, A. R.; Woolley, G. A. *J. Am. Chem. Soc.* 2009, *131*, 2283–2289.

[37] Lubbe, A. S.; Szymanski, W.; Feringa, B. L. *Chem. Soc. Rev.* 2017, *46*, 1052-1079.

[38] Wang, F.; Liu, X.; Willner, I. *Angew. Chem. Int. Ed.* 2015, *54*, 1098-129.

[39] Liu, Q.; Deiters, A. *Acc. Chem. Res.* 2014, *47*, 45-55.

[40] Szymański, W.; Beierle, J. M.; Kistemaker, H. A. V.; Velema, W. A.; Feringa, B. L. *Chem. Rev.* 2013, *113*, 6114-6178.

[41] Li, S.; Jiang, X.; Zheng, R.; Zuo, S.; Zhao, L.; Fan, G.; Fan, J.; Liao, Y.; Yu, X.; Cheng, H. *Chem. Commun.* 2018, *54*, 7983-7986.

[42] Nehls, E. M.; Rosales, A. M.; Anseth, K. S. *J. Mat. Chem. B* 2016, *4*, 1035-1039.

[43] Lutz, T.; Wein, T.; Höfner, G.; Pabel, J.; Eder, M.; Dine, J.; Wanner, K. T. *J. Med. Chem.* 2018, *61*, 6211-6235.

[44] Tian, X.; Zhang, C.; Xu, Q.; Li, Z.; Shao, X. *Org. Biomol. Chem.* 2017, *15*, 3320-3323.

[45] Runtsch, L. S.; Barber, D. M.; Mayer, P.; Groll, M.; Trauner, D.; Broichhagen, J. *Beilstein J. Org. Chem.* 2015, *11*, 1129-1135.

[46] Ferreira, R.; Nilsson, J. R.; Solano, C.; Andréasson, J.; Grøtli, M. *Sci. Rep.* 2015, *5*, 9769.

[47] Broichhagen, J.; Frank, J. A.; Trauner, D. *Acc. Chem. Res.* 2015, *48*, 1947-1960.

[48] Borowiak, M.; Nahaboo, W.; Reynders, M.; Nekolla, K.; Jalinot, P.; Hasserodt, J.; Rehberg, M.; Delattre, M.; Zahler, S.; Vollmar, A.; Trauner, D.; Thorn-Seshold, O. *Cell* 2015, *162*, 403-411.

[49] Wendler, T.; Schütt, C.; Näther, C.; Herges, R. *J. Org. Chem.* 2012, *77*, 3284-3287.

[50] Weston, C. E.; Richardson, R. D.; Haycock, P. R.; White, A. J. P.; Fuchter, M. J. *J. Am. Chem. Soc.* 2014, *136*, 11878-11881.

[51] Leuchs, D.; *Azothiazole Dyes*, FR 1379349 19640120, 1963.

[52] Van de Sande, C. C.; Callant, P. R.; Van den Bergh, A. M.; Agfa-Gevaert N. V. *Cyan Dye-releasing Compounds for Use in the Production of Diffusion-transfer Color Images*, EP 216957, 1987.

[53] Kajikawa, K.; Senda, K.; Masunaka, S.; Hirose, K.; Yamamoto, T.; Ichikawa, A.; Ozeki, T. *Dyeing Polypropylene Fiber*, JP 45014153, 1970.

[54] Gaffer, H. E.; Khalifa, M. E. *Molecules* 2015, *20*, 21982-21991.

[55] Seto, N.; Kamio, T.; Hiyoshi, H. *Isothiazolylazophenol Dye and a Color Photographic Material Containing the Precursor of the Magenta Dye*, JP 09292684, 1997.

[56] Sivik, M. R.; Sadlowski, E. S.; Lim, M.; Javet, M.; Pasquier, C.; Loughnane, B. J.; Cummings, M. D.; Morrall, D. D. A.; Kott, K. L.; Baker, K. H.; Torres, E.; Keiken, L. D.; Moore, P. D.; Valenti, D. J.; Mahaffey, R. L. *2. Laundry Care Compositions with Thiazolium Dyes*, US 20140082851, 2014.

[57] Ribka, J.; Tappe, H.; Roth, K.; Weyer, H. J.; Cassella A.-G. *3. Reserve Effect in Polyester-cellulose-mixed Fiber Textiles*. DE 1978-2856283, 1978.

[58] Zhukova, N. A.; Serova, V. N.; Mamedov, V. A. *Izv. Vyssh. Uchebn. Zaved., Khim. Khim. Tekhnol.* 2006, *49*, 65-69.

[59] Serova, V. N.; Zhukova, N. A.; Mamedov, V. A. *Russ. J. Appl. Chem.* 2006, *79*, 637-641.

[60] Bulka, E.; Beyer, H.; Brandenburg, G. *Chem. Ber.* 1959, *92*, 1447-56.

[61] Weaver, M. A.; Wallace, D. J. *Thiazolylazotetrahydroquinoline Compounds*, US 3787388, 1974.

[62] Eliu, V. P.; Sauder, T.; Kauffmann, D.; Froehling, B. *2. 1, 3-disubstituted-2-[(2-amino-5-acylaminophenyl)azo]imidazolium Salts for Hair Dyes*, GB 2412661, 2005.

[63] Balli, H.; Chaquime, L.; Hellrung, B.; Misun, M. *Helv. Chim. Acta* 1994, *77*, 216-230.

[64] Jablonski, J.; Muszynski, M.; Scibisz, H. *Azo Dye Mixtures for Coloring Polyester Fabrics Yellow*, PL 150641, 1990.

[65] Koenigs, E.; Freigang, W.; Lobmayer, G.; Zscharn, A. *Ber. Dtsch. Chem. Ges. B* 1926, *59B*, 321.

[66] Grabener, E. *Internist* 1960, *1*, 179-81.
[67] Blaise, C.; David, H.; L'Oreal. *4. Process of Coloring of Keratinous Fibers Implementing at Least a Particular 2- or 4-azopyridinium Dye and at Least a Fluorescent Dye*, WO 2018115155, 2018.
[68] Crabtree, A.; Parton, B. *Process for the manufacture of azo and hydrazo dyes and certain dyes so obtained,* GB 2032448, 1980.
[69] Andrew, H. F.; Poole, V. D. *Chloropyrimidinyl reactive dyes,* US 3097910, 1963.
[70] Balli, H.; Ritter, H. *Dyes Pigm.* 1981, *2*, 93-124.
[71] Basaif, S. A.; Hassan, M. A.; Gobouri, A. A. *Dyes Pigm.* 2006, *72*, 387-391.
[72] Ben-Bassat, A.; Dershowitz, S. *J. Soc. Dyers Colour.* 1979, *95*, 143-7.
[73] Benz, J.; Schweizer, A.; Wehrli, W. *Reactive dyes containing alkylidenehydrazino groups,* BE 635676, 1963.
[74] Brouard, C. M. H. E.; Leroy, J. M. L.; Stiot, J. P. H. *Phenylazocinnamonitriles,* DE 2111734, 1971.
[75] Casoni, D. D. M. *Boll. Sci. Fac. Chim. Ind. Bologna* 1956, *14*, 22-30.
[76] El-Bindary, A. A.; El-Sonbati, A. Z.; Al-Sarawy, A. A.; Mohamed, K. S.; Farid, M. A. *Spectrochima Acta Part A* 2015, *136*, 1842-1849.
[77] El-Bindary, A. A.; El-Sonbati, A. Z.; Farid, M. A.; Al-Sarawy, A. A.; Mohamed, K. S. *Spectrochima Acta Part A* 2015, *136*, 1842-1849.
[78] El-Hefnawy, G. B.; El-Borai, M. A.; Aly, E. S. A.; Gabr, A. A. *Indian J. Fibre Text. Res.* 1992, *17*, 87-91.
[79] El-Sayed, G. M.; El-Sabbagh, I. A.; Ismail, L. F. M.; Imam, A. G.; El-Mariah, A. A. R. *Zh. Anal. Khim.* 1990, *45*, 2272-2274.
[80] El-Sayed, G. M.; El-Sabbagh, I. A.; Ismail, L. F. M.; Imam, A. G.; El-Mariah, A. A. R. *Sci. Phys. Sci.* 1991, *3*, 101-105.
[81] El-Sayed, G. M.; El-Sabbagh, I. A.; Ismail, L. F. M.; Imam, A. G.; El-Mariah, O. A. R. *Zh. Anal. Khim.* 1991, *46*, 2209-2213.
[82] Fahmy, S. M.; Badran, A. H.; Elnagdi, M. H. *J. Chem. Technol. Biotechnol.* 1980, *30*, 390-395.

[83] Green, J. A., II; Kalenda, N. W. *Photographic products and processes employing nondiffusible pyridylazopyrazole or pyrimidylazopyrazole dye-releasing compounds,* US 4148641, 1979.

[84] Haase, J. R.; Eldredge, C. H.; Landholm, R. A. *Yellow azopyrazoline dye releasing redox compounds for photographic color transfer,* US 4013633, 1977.

[85] Hegar, G.; Back, G. *Azo dyes,* DE 2349752, 1974.

[86] Idelson, E. M. *Light sensitive recording material for the dye developer diffusion process,* DE 1572029, 1970.

[87] Italia, A. *Tinctoria* 1997, *94*, 48-56.

[88] Kobrakov, K. I.; Feoktistov, M. K.; Stankevich, G. S.; Dmitrieva, M. B.; Brodskii, I. I.; Begunov, R. S. *Method for production of heat-resistant perinone dyes,* RU 2303614, 2007.

[89] Konstantinov, V. N.; Al'perovich, A. P.; Al'perovich, M. A.; Zil'berman, I. I.; Abramenko, P. I.; Gol'tsev, A. N.; Shub, M. A. *Composition for making a color adjustment optical film filter*, SU 1068458, 1984.

[90] Kovalchukova, O. V.; Abdulilla Abbas, A. T. R.; Strashnova, S. B.; Strashnov, P. V. *Rev. Inorg. Chem.* 2018, *38*, 87-101.

[91] Kuznetsov, D. N.; Ruchkina, A. G.; Kobrakov, K. I. *Chem. Heterocycl. Compd.* 2011, *47*, 441-447.

[92] Metwally, M. A.; Bondock, S.; El-Desouky, E.-S. I.; Abdou, M. M. *Color. Technol.* 2013, *129*, 418-424.

[93] Metwally, M. A.; Bondock, S. A.; El-Desouky, S. I.; Abdou, M. M. *J. Korean Chem. Soc.* 2012, *56*, 82-91.

[94] Moussa, E. A.; El-Sabbagh, I. A.; Sallam, H. B.; Moustafa, O. I.; El-Mariah, A. A. R. *Orient. J. Chem.* 1990, *6*, 258-63.

[95] Popa, G.; Baiulescu, G.; Moldoveanu, S. *Acad. Rep. Populate Romine, Studii Cercetari Chim [Acad. Rep. Populate Romine, Chim Research].* 1963, *11*, 287-90.

[96] Reber, L.; Doerffeldt, J.; Schulz, E.; Starke, G. *Dyeing of synthetic high polymers, especially polyamides, in the manufacturing process*, DD 115724, 1975.

[97] Roos, H.; *General Aniline & Film Corp*. 1940.

[98] Skoweranda, J.; Bukowska-Strzyzewska, M.; Strzyzewski, W. *J. Chem. Crystallogr.* 1994, *24*, 517-520.

[99] Wawrzyniak, A. W.; Oczkowski, M. *Przegl. Wlok.* 1971, *25*, 647-653.

[100] Le Fèvre, R. J. W.; Worth, C. V. *J. Chem. Soc.* 1951, 1814-1817.

[101] Campbell, N.; Henderson, A. W.; Taylor, D. *J. Chem. Soc.* 1953, 1281-1285.

[102] Dougan, S. J.; Melchart, M.; Habtemariam, A.; Parsons, S.; Sadler, P. J. *Inorg. Chem.* 2006, *45*, 10882-10894.

[103] Ortiz, M.; Torréns, M.; Mola, J. L.; Ortiz, P. J.; Fragoso, A.; Díaz, A.; Cao, R.; Prados, P.; de Mendoza, J.; Otero, A.; Antiñolo, A.; Lara, A. *Dalton Trans.* 2008, 3559-3566.

[104] Manojkumar, P.; Ravi, T. K.; Gopalakrishnan, S. *Eur. J. Med. Chem.* 2009, *44*, 4690-4694.

[105] Towns, A. D. *Dyes Pigm.* 1999, *42*, 3-28.

[106] Weaver, M. A.; Shuttleworth, L. *Dyes Pigm.* 1982, *3*, 81-121.

[107] Byabartta, P. *Int. J. Curr. Res. Chem. Pharm. Sci.* 2015, *2*, 75-81.

[108] Pandey, P.; Maheshwari, N.; Dave, A. *J. Liq. Chromatogr.* 1991, *14*, 3311-15.

[109] Al-Shiekh, M. A.; Salah El-Din, A. M.; Hafez, E. A.; Elnagdi, M. H. *J. Heterocycl. Chem.* 2009, *41*, 647-654

[110] Khound, S.; Das, P. J. *Indian J. Chem., Sect. B: Org. Chem. Incl. Med. Chem.* 1998, *37B*, 155-157.

[111] Jain, R.; Agarwal, D. D.; Goyal, R. N. *Fresenius. Anal. Chem.* 1979, *298*, 44.

[112] Sharma, P.; Kumar, A.; Upadhyay, S.; Singh, J.; Sahu, V. *Med. Chem. Res.* 2010, *19*, 589-602.

[113] Marten, J.; Seichter, W.; Weber, E. *Z. Anorg. Allg. Chem.* 2005, *631*, 869-877.

[114] Kunitomo, J.; Yoshikawa, M.; Fushimi, M.; Kawada, A.; Quinn, J. F.; Oki, H.; Kokubo, H.; Kondo, M.; Nakashima, K.; Kamiguchi, N.; Suzuki, K.; Kimura, H.; Taniguchi, T. *J. Med. Chem.* 2014, *57*, 9627-9643.

[115] Kale, P. D. *J. Chem. Pharm. Res.* 2013, *5*, 130-134.

[116] Wang, Z.; Yin, Z.; Zhu, F.; Li, Y.; Wu, X.-F. *Chem Cat Chem* 2017, *9*, 3637-3640.

[117] Iranpoor, N.; Firouzabadi, H.; Khalili, D. *Org. Biomol. Chem.* 2010, *8*, 4436-4443.

[118] Jaunet-Lahary, T.; Chantzis, A.; Chen, K. J.; Laurent, A. D.; Jacquemin, D. *J. Phys. Chem. C* 2014, *118*, 28831-28841.

[119] Garci, A.; Dobrov, A. A.; Riedel, T.; Orhan, E.; Dyson, P. J.; Arion, V. B.; Therrien, B. *Organometallics* 2014, *33*, 3813-3822.

[120] Fu, Y.; Romero, M. J.; Salassa, L.; Cheng, X.; Habtemariam, A.; Clarkson, G. J.; Prokes, I.; Rodger, A.; Costantini, G.; Sadler, P. J. *Angew. Chem.Int. Ed.* 2016, *55*, 8909-8912.

[121] Zhou, Q.-X.; Zheng, Y.; Wang, T.-J.; Chen, Y.-J.; Li, K.; Zhang, Y.-Y.; Li, C.; Hou, Y.-J.; Wang, X.-S. *Chem. Commun.* 2015, *51*, 10684-10686.

[122] Rudenko, A. E.; Clayman, N. E.; MacLaren, J. K.; Waymouth, R. M. *ChemistrySelect* 2016, *1*, 3491-3496.

[123] Eskandari, A.; Kundu, A.; Lu, C.; Ghosh, S.; Suntharalingam, K. *Dalton Trans.* 2018, *47*, 5755-5763.

[124] Telleria, A.; van Leeuwen, P. W. N. M.; Freixa, Z. *Dalton Trans.* 2017, *46*, 3569-3578.

[125] Ghebreyessus, K.; Cooper, S. M. *Organometallics* 2017, *36*, 3360-3370.

[126] Gupta, P.; Karothu, D. P.; Ahmed, E.; Naumov, P.; Nath, N. K. *Angew. Chem. Int. Ed.* 2018, *57*, 8498-8502.

[127] Hutchins, K. M.; Unruh, D. K.; Verdu, F. A.; Groeneman, R. H. *Cryst. Growth Des.* 2018, *18*, 566-570.

[128] Santana, C. M.; Reinheimer, E. W.; Krueger, H. R.; MacGillivray, L. R.; Groeneman, R. H. *Cryst. Growth Des.* 2017, *17*, 2054-2058

[129] Drath, O.; Gable, R. W.; Moubaraki, B.; Murray, K. S.; Poneti, G.; Sorace, L.; Boskovic, C. *Inorg. Chem.* 2016, *55*, 4141-4151.

[130] Perlovich, G. L. *CrystEngComm* 2015, *17*, 7019-7028.

[131] Ravat, P.; SeethaLekshmi, S.; Biswas, S. N.; Nandy, P.; Varughese, S. *Cryst. Growth Des.* 2015, *15*, 2389-2401.

[132] Bushuyev, O. S.; Corkery, T. C.; Barrett, C. J.; Friščić, T. *Chem. Sci.* 2014, *5*, 3158-3164.

[133] Aakeröy, C. B.; Panikkattu, S.; DeHaven, B.; Desper, J. *Cryst Eng Comm* 2013, *15*, 463-470.

[134] Wang, L.-C.; Sun, J.; Huang, Z.-T.; Zheng, Q.-Y. *Cryst. Growth Des.* 2013, *13*, 1-5

[135] Zaltariov, M.-F.; Cazacu, M.; Shova, S.; Vlad, A.; Stoica, I.; Hamciuc, E. *J. Polym. Sci., Part A: Polym. Chem.* 2012, *50*, 3775-3787.

[136] Wasilewska, A.; Milewska, M. J.; Gdaniec, M.; Połoński, T. *Tetrahedron: Asymmetry* 2009, *20*, 1472-1475.

[137] Zhang, J.; Ye, L.; Wu, L.-X. *Acta Cryst. C* 2005, *61*, 38-40.

[138] Rudenko, A. E.; Clayman, N. E.; Walker, K. L.; Maclaren, J. K.; Zimmerman, P. M.; Waymouth, R. M. *J. Am. Chem. Soc.* 2018, *140*, 11408-11415.

[139] Singh, S. *ChemistrySelect* 2018, *3*, 6786-6790.

[140] Loya, J. D.; Qiu, J.; Unruh, D. K.; Cozzolino, A. F.; Hutchins, K. M. *Cryst. Growth Des.* 2018, *18*, 4838-4843.

[141] Waitschat, S.; Reinsch, H.; Arpacioglu, M.; Stock, N. *CrystEng Comm.* 2018, *20*, 5108-5111.

[142] Zhang, Z.; Cui, X.; Yang, L.; Cui, J.; Bao, Z.; Yang, Q.; Xing, H. *Ind. Eng. Chem. Res* 2018, *57*, 7266-7274.

[143] Cui, J.; Du, X.; Li, C. *Inorg. Nano-Met. Chem.* 2018, *48*, 44-48.

[144] Senthilkumar, S.; Goswami, R.; Obasi, N. L.; Neogi, S. *ACS Sustainable Chem. Eng.* 2017, *5*, 11307-11315.

[145] Venkataramani, S.; Jana, U.; Dommaschk, M.; Sönnichsen, F. D.; Tuczek, F.; Herges, R. *Science* 2011, *331*, 445-448.

[146] Herges, R. *Nachrichten aus der Chemie [News from the Chemistry]* 2011, *59*, 817-821.

[147] Dommaschk, M.; Peters, M.; Gutzeit, F.; Schütt, C.; Näther, C.; Sönnichsen, F. D.; Tiwari, S.; Riedel, C.; Boretius, S.; Herges, R. *J. Am. Chem. Soc.* 2015, *137*, 7552-7555.

[148] Gómez-Santacana, X.; Pittolo, S.; Rovira, X.; Lopez, M.; Zussy, C.; Dalton, J. A. R.; Faucherre, A.; Jopling, C.; Pin, J.-P.; Ciruela, F.;

Goudet, C.; Giraldo, J.; Gorostiza, P.; Llebaria, A. *ACS Cent. Sci.* 2017, *3*, 81-91.

[149] Simeth, N. A.; Crespi, S.; Fagnoni, M.; König, B. *J. Am. Chem. Soc.* 2018, *140*, 2940-2946.

[150] Calbo, J.; Weston, C. E.; White, A. J. P.; Rzepa, H. S.; Contreras-García, J.; Fuchter, M. J. *J. Am. Chem. Soc.* 2017, *139*, 1261-1274.

[151] Devi, S.; Saraswat, M.; Grewal, S.; Venkataramani, S. *J. Org. Chem.* 2018, *83*, 4307-4322.

[152] Stricker, L.; Böckmann, M.; Kirse, T. M.; Doltsinis, N. L.; Ravoo, B. J. *Chem. Eur. J.* 2018, *24*, 8639-8647.

[153] Xu, Y.; Gao, C.; Andréasson, J.; Grøtli, M. *Org. Lett.* 2018, *20*, 4875-4879.

[154] Diener, H.; Zollinger, H. *Can. J. Chem.* 1986, *64*, 1102-1107.

[155] Diener, H.; Güleç, B.; Skrabal, P.; Zollinger, H. *Helv. Chim. Acta* 1989, *72*, 800-805.

[156] Goerdeler, J.; Haubrich, H. *Chem. Ber.* 1960, *93*, 397-405.

[157] Brown, R. D.; Duffin, H. C.; Maynard, J. C.; Ridd, J. H. *J. Chem. Soc.* 1953, 3937-3939.

[158] Macháčeka, V; Kořínek, J; Kreuzigovváb, D; Štěrbaa, V. *Collect. Czech. Chem. Commun.* 1985, *50*, 658-674.

[159] Butler, A. R.; Pogorzelec, P.; Shepherd, P. T. *J. Chem. Soc., Perkin Trans. 2* 1977, 1452-1457.

[160] Anderson, L. M.; Butler, A. R.; Glidewell, C.; Hart, D.; Isaacs, N. *J. Chem. Soc., Perkin Trans. 2* 1989, 2055-2058.

[161] Dommaschk, M.; Schütt, C.; Venkataramani, S.; Jana, U.; Näther, C.; Sönnichsen, F. D.; Herges, R. *Dalton Trans.* 2014, *43*, 17395-17405.

[162] Heitmann, G.; Schütt, C.; Herges, R. *Eur. J. Org. Chem.* 2016, *2016*, 3817-3823.

[163] Kolarski, D.; Szymanski, W.; Feringa, B. L. *Org. Lett.* 2017, *19*, 5090-5093.

[164] Anderson, L. M.; Butler, A. R.; Glidewell, C.; Hart, D.; Isaacs, N. *J. Chem. Soc., Perkin Trans. 2* 1989, 2055-2058.

[165] Jacquemin, D. *Int. J. Quantum Chem.* 2012, *112*, 2043-2050.

[166] Yin, T.-T.; Zhao, Z.-X.; Zhang, H.-X. *New J. Chem.* 2017, *41*, 1659-1669.

[167] Voigt, M.; Hallensleben, M. L.; Toppare, L. *Synthetic Metals* 1993, *55*, 1067-1072.

[168] Bottaro, J. C.; Schmitt, R. J.; Penwell, P. E. *N, N'-Azobisnitroazoles and analogs as detonators and inflators for occupant restraint devices and airbags*, US 5889161, 1999.

[169] Ballard, H. H. *Tetrahedron Lett.* 1989, *30*, 2345-2348.

[170] Bartle, M. G.; Gore, S. T.; Mackie, R. K.; Mhatre, S.; Tedder, J. M. *J. Chem. Soc., Perkin Trans. 1* 1978, 401-406.

[171] Das, D.; Agarkar, H. *ACS Omega* 2018, *3*, 7630-7638.

[172] Faessinger, R. W.; Brown, E. V. *J. Am. Chem. Soc.* 1951, *73*, 4606-4608.

[173] Korom, S.; Ballester, P. *J. Am. Chem. Soc.* 2017, *139*, 12109-12112.

[174] Waldie, K. M.; Ramakrishnan, S.; Kim, S.-K.; Maclaren, J. K.; Chidsey, C. E. D.; Waymouth, R. M. *J. Am. Chem. Soc.* 2017, *139*, 4540-4550.

[175] Brown, E. V.; Granneman, G. R. *J. Am. Chem. Soc.* 1975, *97*, 621-627.

[176] Hutchins, K. M.; Yelgaonkar, S. P.; Harris-Conway, B. L.; Reinheimer, E. W.; MacGillivray, L. R.; Groeneman, R. H. *Supramol. Chem.* 2018, *30*, 533-539.

[177] Thies, S.; Sell, H.; Bornholdt, C.; Schuett, C.; Koehler, F.; Tuczek, F.; Herges, R. *Chem. Eur. J.* 2012, *18*, 16358-16368.

[178] Bannwarth, A.; Schmidt, S. O.; Peters, G.; Soennichsen, F. D.; Thimm, W.; Herges, R.; Tuczek, F. *Eur. J. Inorg. Chem.* 2012, *2012*, 2776-2783.

[179] Wang, Y.; Ge, X.; Schull, G.; Berndt, R.; Bornholdt, C.; Koehler, F.; Herges, R. *J. Am. Chem. Soc.* 2008, *130*, 4218-4219.

[180] Wang, Y.; Ge, X.; Schull, G.; Berndt, R.; Tang, H.; Bornholdt, C.; Koehler, F.; Herges, R. *J. Am. Chem. Soc.* 2010, *132*, 1196-1197.

[181] Chowdhury, B.; Naskar, K.; Mallick, D.; Sarkar, K. K.; Sen, C.; Sinha, C. *Inorg. Chim. Acta* 2018, *483*, 87-92.

[182] Misra, T. K.; Das, D.; Sinha, C. *Polyhedron* 1997, *16*, 4163-4170.

[183] Wang, Y.-T.; Liu, X.-Y.; Cui, G.; Fang, W.-H.; Thiel, W. *Angew. Chem. Int. Ed.* 2016, *55*, 14009-14013.

[184] Boulton, A. J.; Katritzky, A. R.; Hamid, A. M. *J. Chem. Soc. C* 1967, 2005-2007.

[185] Bustos, C.; Molins, E.; Carcamo, J.-G.; Aguilar, M. N.; Sanchez, C.; Moreno-Villoslada, I.; Nishide, H.; Mesias-Salazar, A.; Zarate, X.; Schott, E. *New J. Chem.* 2015, *39*, 4295-4307.

[186] Garg, H. G.; Singh, P. P. *J. Med. Chem.* 1970, *13*, 1250-1251.

[187] Schuett, C.; Heitmann, G.; Wendler, T.; Krahwinkel, B.; Herges, R. *J. Org. Chem.* 2016, *81*, 1206-1215.

[188] Sagebiel, S.; Stricker, L.; Engel, S.; Ravoo, B. J. *Chem. Commun.* 2017, *53*, 9296-9299.

[189] Al-Shiekh, M. A.; Salah El-Din, A. M.; Hafez, E. A.; Elnagdi, M. H. *J. Heterocycl. Chem.* 2004, *41*, 647-654.

[190] Kulaeva, L. N.; Grabenko, A. D.; Pel'kis, P. S. *Khim. Geterotsikl. Soedin.* 1978, 909-911.

[191] Whittaker, N.; Jones, T. S. G. *J. Chem. Soc.* 1951, 1565-1570.

[192] Santra, P. K.; Das, D.; Misra, T. K.; Roy, R.; Sinha, C.; Peng, S.-M. *Polyhedron* 1999, *18*, 1909-1915.

[193] Senapoti, S.; Jasimuddin, S.; Mostafa, G.; Lu, T. H.; Sinha, C. *Polyhedron* 2006, *25*, 1571-1578.

[194] Shneshil, M. K.; Ibraheem, T. K.; Jamal, M. Y. *J. Appl. Chem.* 2014, *3*, 2312-2317.

[195] Coelho, P. J.; Castro, M. C. R.; Fernandes, S. S. M.; Fonseca, A. M. C.; Raposo, M. M. M. *Tetrahedron Lett.* 2012, *53*, 4502-4506.

[196] Broichhagen, J.; Frank, J. A.; Johnston, N. R.; Mitchell, R. K.; Šmid, K.; Marchetti, P.; Bugliani, M.; Rutter, G. A.; Trauner, D.; Hodson, D. J. *Chem. Commun.* 2015, *51*, 6018-6021.

[197] Inoue, Y.; Kamo, Y.; Togo, M.; Hsu, Y. C.; Keumi, T.; Kitajima, H. *Chem. Lett.* 1981, 1733-1736.

[198] Li, Y.-C.; Qi, C.; Li, S.-H.; Zhang, H.-J.; Sun, C.-H.; Yu, Y.-Z.; Pang, S.-P. *J. Am. Chem. Soc.* 2010, *132*, 12172-12173.

[199] Schultheiss, N. C.; Holtsclaw, J. *Energetic cocrystals for treatment of a subterranean formation,* WO 2015030730, 2015.

[200] Klapoetke, T. M.; Piercey, D. G. *Inorg. Chem.* 2011, *50*, 2732-2734.

[201] Qi, C.; Zhang, R.-B.; Pang, S.-P. *RSC Adv.* 2013, *3*, 17741-17748.

[202] Hiskey, M. A.; Goldman, N.; Stine, J. R. *J. Energ. Mater.* 1998, *16*, 119-127.

[203] Jiao, B.; Ren, Y.; Wang, W. *J. Mol. Struct.* 2018, *1171*, 150-154.

[204] Cohen, A.; Yang, Y.; Yan, Q.-L.; Shlomovich, A.; Petrutik, N.; Burstein, L.; Pang, S.-P.; Gozin, M. *Chem. Mater.* 2016, *28*, 6118-6126.

[205] Li, S. H.; Pang, S. P.; Li, X. T.; Yu, Y. Z.; Zhao, X. Q. *Chin. Chem. Lett.* 2007, *18*, 1176-1178.

[206] Qi, C.; Li, S.-H.; Li, Y.-C.; Wang, Y.; Chen, X.-K.; Pang, S.-P. *J. Mater. Chem.* 2011, *21*, 3221-3225.

[207] Zhang, J.; Du, Y.; Dong, K.; Su, H.; Zhang, S.; Li, S.; Pang, S. *Chem. Mater.* 2016, *28*, 1472-1480.

[208] Liu, X.-M.; Wang, B.-Y.; Xue, W.; Xie, L.-H.; Zhang, W.-X.; Cheng, X.-N.; Chen, X.-M. *Dalton Trans.* 2012, *41*, 13741-13746.

[209] Khalili, D.; Iranpoor, N.; Firouzabadi, H. *J. Sulfur Chem.* 2015, *36*, 544-555.

[210] Markees, D.; Kellerhals, M.; Erlenmeyer, H. *Helv. Chim. Acta* 1947, *30*, 304-306.

[211] Ruggli, P.; von Wartburg, A.; Erlenmeyer, H. *Helv. Chim. Acta* 1947, *30*, 348-353.

[212] Glover, M.; Huddleston, P. R.; Wood, M. L. *J. Chem. Res.* 2013, *37*, 43-44.

[213] Okumura, S.; Lin, C.-H.; Takeda, Y.; Minakata, S. *J. Org. Chem.* 2013, *78*, 12090-12105.

[214] Sahoo, M. K.; Saravanakumar, K.; Jaiswal, G.; Balaraman, E. *ACS Catal.* 2018, *8*, 7727-7733.

[215] Wang, L.; Ishida, A.; Hashidoko, Y.; Hashimoto, M. *Angew. Chem. Int. Ed.* 2017, *56*, 870-873.

[216] He, N.; Zhang, Y.; Liu, R.; Guo, R.; Suo, Z. *J. Therm. Anal. Calorim.* 2017, *129*, 515-520.

[217] Hermann, T. S.; Klapoetke, T. M.; Krumm, B.; Stierstorfer, J. *J. Heterocycl. Chem.* 2018, *55*, 852-862.

[218] Sheremetev, A. B.; Palysaeva, N. V.; Struchkova, M. I.; Suponitsky, K. Y. *Mendeleev Commun.* 2012, *22*, 302-304.

[219] Wei, X.; Zhang, A.; Ma, Y.; Xue, X.; Zhou, J.; Zhu, Y.; Zhang, C. *CrystEngComm* 2015, *17*, 9037-9047.

[220] Zhang, J.; Shreeve, J. n. M. *J. Phys. Chem. C* 2015, *119*, 12887-12895.

[221] Baldwin, D. A.; Lever, A. B. P.; Parish, R. V. *Inorg. Chem.* 1969, *8*, 107-115.

[222] Kottelat, E.; Ruggi, A.; Zobi, F. *Dalton Trans.* 2016, *45*, 6920-6927.

[223] Maity, S.; Kundu, S.; Weyhermuller, T.; Ghosh, P. *Inorg. Chem.* 2015, *54*, 1300-1313.

[224] Sarkar, B.; Patra, S.; Fiedler, J.; Sunoj, R. B.; Janardanan, D.; Lahiri, G. K.; Kaim, W. *J. Am. Chem. Soc.* 2008, *130*, 3532-3542.

[225] Singh, S.; Chauhan, P.; Ravi, M.; Taneja, I.; Wahajuddin; Yadav, P. P. *RSC Adv.*2015, *5*, 61876-61880.

[226] Chand, S.; Pal, A.; Pal, S. C.; Das, M. C. *Eur. J. Inorg. Chem.* 2018, *2018*, 2785-2792.

[227] Friedl, F. *Monatsh. Chem.* 1913, *34*, 759-767.

[228] Schoenfeld, S.; Lochenie, C.; Thoma, P.; Weber, B. *CrystEngComm* 2015, *17*, 5389-5395.

[229] Thies, S.; Sell, H.; Schuett, C.; Bornholdt, C.; Naether, C.; Tuczek, F.; Herges, R. *J. Am. Chem. Soc.* 2011, *133*, 16243-16250.

[230] Boyland, E.; Sims, P. *J. Chem. Soc.* 1958, 4198-4199.

[231] Jiang, M.; Li, B.; Cui, X.; Yang, Q.; Bao, Z.; Yang, Y.; Wu, H.; Zhou, W.; Chen, B.; Xing, H. *ACS Appl. Mater. Interfaces* 2018, *10*, 16628-16635.

[232] Liu, G.; Sheng, J.; Wu, H.; Yang, C.; Yang, G.; Li, Y.; Ganguly, R.; Zhu, L.; Zhao, Y. *J. Am. Chem. Soc.* 2018, *140*, 6467-6473.

[233] Sharma, V.; De, D.; Pal, S.; Saha, P.; Bharadwaj, P. K. *Inorg. Chem.* 2017, *56*, 8847-8855.

[234] Chavez, D. E.; Hiskey, M. A.; Gilardi, R. D. *Angew. Chem. Int. Ed.* 2000, *39*, 1791-1793.

[235] Jafari, M.; Keshavarz, M. H.; Noorbala, M. R.; Kamalvand, M. *Chemistry Select* 2016, *1*, 5286-5296.

[236] Keshavarz, M. H.; Esmaeilpour, K.; Oftadeh, M.; Abadi, Y. H. *RSC Adv.* 2015, *5*, 87392-87399.

[237] Smeu, M.; Zahid, F.; Ji, W.; Guo, H.; Jaidann, M.; Abou-Rachid, H. *J. Phys. Chem. C* 2011, *115*, 10985-10989.

[238] Koppes, W. M.; Moran, J. S.; Oxley, J. C.; Smith, J. L. *Tetrahedron Lett.* 2008, *49*, 3234-3237.

[239] Liu, W.; Lin, Q.-H.; Yang, Y.-Z.; Zhang, X.-J.; Li, Y.-C.; Lin, Z.-H.; Pang, S.-P. *Chem. Asian J.* 2014, *9*, 479-486.

[240] Oxley, J. C.; Smith, J. L.; Moran, J. S. *J. Energ. Mater.* 2009, *27*, 63-93.

[241] Cankar, P.; Frisova, I.; Krystof, V.; Lenobel, R.; Slouka, J.; Strnad, M.; Fisher, P. M. *4-Arylazo-3, 5-diaminopyrazole compounds and their preparation, pharmaceutical compositions, inhibition of cyclin-dependent kinases and treatment of proliferative and viral disorders*, US 20080312238, 2006, 101.

[242] Mathur, T.; Dinda, J.; Datta, P.; Mostafa, G.; Lu, T. H.; Sinha, C. *Polyhedron* 2006, *25*, 2503-2512.

[243] Mathur, T.; Jasimuddin, S.; Milton, H.; Woollins, J. D.; Sinha, C. *Inorg. Chim. Acta* 2004, *357*, 3503-3509.

[244] Kitagawa, T.; Iwakura, K.; Hirai, E. *Chem. Pharm. Bull.* 1981, *29*, 2303-2307.

[245] Jorda, R.; Navratilova, J.; Huskova, Z.; Schuetznerova, E.; Cankar, P.; Strnad, M.; Krystof, V. *Chem. Biol. Drug Des.* 2014, *84*, 402-408.

[246] Jorda, R.; Schutznerova, E.; Cankar, P.; Brychtova, V.; Navratilova, J.; Krystof, V. *Bioorg. Med. Chem.* 2015, *23*, 1975-1981.

[247] Zhang, Z.; Daynard, T. S.; Chafeev, M. A.; Wang, S.; Chopiuk, G. B.; Sviridov, S. V. *Preparation of hydrazonodiaminopyrazoles as integrin-linked kinase inhibitors with antiproliferative activity US 20030060453*, 2003.

[248] Kirk, W., Jr.; Johnson, J. R.; Blomquist, A. T. *J. Org. Chem.* 1943, *8*, 557-563.

[249] Iranpoor, N.; Firouzabadi, H.; Khalili, D. *Bull. Chem. Soc. Jpn.* 2010, *83*, 923-934.

[250] Iranpoor, N.; Firouzabadi, H.; Khalili, D. *Tetrahedron Lett.* 2012, *53*, 6913-6915.

[251] Iranpoor, N.; Firouzabadi, H.; Shahin, R.; Khalili, D. *Synth. Commun.* 2012, *42*, 2040-2047.

[252] Khalili, D.; Iranpoor, N.; Firouzabadi, H. *J. Sulfur Chem.* 2015, *36*, 544-555.

[253] Patel, S.; Mishra, B. K. *Tetrahedron Lett.* 2004, *45*, 1371-1372.

[254] Pozharskii, A. F.; Zvezdina, E. A.; Simonov, A. M. *Tetrahedron Lett.* 1967, 2219-2222.

[255] Wang, M.; Funabiki, K.; Matsui, M. *Dyes Pigm.* 2003, *57*, 77-86.

[256] Nay, B.; Scriven, E. F. V.; Suschitzky, H.; Thomas, D. R. *J. Chem. Soc., Perkin Trans. 1* 1980, 611-613.

[257] Badger, G. M.; Buttery, R. G. *J. Chem. Soc.* 1955, 2816-2818.

[258] Cao, D.; Zhang, Y.; Liu, C.; Wang, B.; Sun, Y.; Abdukadera, A.; Hu, H.; Liu, Q. *Org. Lett.* 2016, *18*, 2000-2003.

[259] Seferoglu, Z.; Ertan, N. *Heteroat. Chem.* 2007, *18*, 622-630.

[260] Kreutzberger, A.; Kalter, P. A. *J. Org. Chem.* 1961, *26*, 3790-3796.

[261] Faustino, H.; Brannigan, C. R.; Reis, L. V.; Santos, P. F.; Almeida, P. *Dyes Pigm.* 2009, *83*, 88-94.

[262] Faustino, H.; El-Shishtawy, R. M.; Reis, L. V.; Santos, P. F.; Almeida, P. *Tetrahedron Lett.* 2008, *49*, 6907-6909.

[263] Hayes, H. T.; Hunter, L. *J. Chem. Soc.* 1941, 1-5.

[264] Binks, J. H.; Ridd, J. H. *J. Chem. Soc.* 1957, 2398-2402.

[265] Vysotskii, Y. B.; Buzykin, B. I.; Bryantsev, V. S. *Chem. Heterocycl. Compd.* 2006, *42*, 331-342.

[266] Petrie, C.; Orme, M. W.; Baindur, N.; Robbins, K. G.; Harris, S. M.; Kontoyianni, M.; Hurley, L. H.; Kerwin, S. M.; Mundy, G. R. *Preparation of (hetero)aromatic compounds for treating bone deficit conditions*, WO 9715308, 1997.

[267] Brown, E. V. *J. Heterocycl. Chem.* 1969, *6*, 571-573.

[268] Yu, X.; Ding, W.; Ge, P.; Wang, S.; Wang, J. *Adv. Synth. Catal.* 2018, *360*, 3150-3156.

[269] Styskala, J.; Lycka, A.; Slouka, J. *J. Heterocycl. Chem.* 2002, *39*, 1305-1308.

[270] Sacmaci, S.; Sacmaci, M. *Appl. Organomet. Chem.* 2017, *31*, 2381-2388.

[271] Semithenko, E. V.; Rort, E. V.; Suboch, G. A. *Izv. Vyssh. Uchebn. Zaved., Khim. Tekhnol.* 2005, *48*, 109-112.

[272] Rayner, C. M.; Graham, M. A. *Sci. Synth.* 2001, *10*, 155-184.

[273] Bailliez, V.; El Kaim, L.; Michaut, V. *Synth. Commun.* 2004, *34*, 109-118.

[274] Razus, A. C.; Birzan, L.; Nae, S.; Razus, S. A.; Cimpeanu, V.; Stanciu, C. *Synth. Commun.* 2002, *32*, 825-837.

[275] Sarma, P. K.; Das, P. J. *Indian J. Chem., Sect. B: Org. Chem. Incl. Med. Chem.* 1995, *34B*, 1092-1094.

[276] Frej, H.; Daszkiewicz, Z.; Kyziol, J. B. *Chem. Pap.* 1990, *44*, 191-199.

[277] Kyziol, J. B.; Ejsmont, K. *Acta Crystallogr., Sect. C: Cryst. Struct. Commun.* 2007, *63*, 77-79.

[278] Gevorgyan, A.; Mkrtchyan, S.; Grigoryan, T.; Iaroshenko, V. O. *ChemPlusChem* 2018, *83*, 375-382.

[279] Fu, J.; Lau, K.; Barra, M. *J. Org. Chem.* 2009, *74*, 1770-1773.

[280] Wang, C.; Chen, H.; Wang, Z.; Chen, J.; Huang, Y. *Angew. Chem. Int. Ed.* 2012, *51*, 7242-7245.

[281] Zhang, Y.; Li, Y.; Zhang, X.; Jiang, X. *Chem. Commun.* 2015, *51*, 941-944.

[282] Zhang, Y.; Hu, H.; Liu, C.-J.; Cao, D.; Wang, B.; Sun, Y.; Abdukader, A. *Asian J. Org. Chem.* 2017, *6*, 102-107.

[283] Wu, Z.; Moore, J. S. *Tetrahedron Lett.* 1994, *35*, 5539-5542.

[284] Petrović, M.; Scarpi, D.; Nieger, M.; Jung, N.; Occhiato, E. G.; Bräse, S. *RSC Adv.* 2017, *7*, 9461-9464.

[285] Gelin, S.; Pollet, P. *J. Heterocycl. Chem.* 1981, *18*, 719-721.

[286] Tanaka, K.; Matsuo, K.; Nakanishi, A.; Jo, M.; Shiota, H.; Yamaguchi, M.; Yoshino, S.; Kawaguchi, K. *Chem. Pharm. Bull.* 1984, *32*, 3291-3298.

[287] Sakamoto, T.; Terao, Y.; Sekiya, M. *Chem. Pharm. Bull.* 1977, *25*, 731-739.

[288] Rueck-Braun, K.; Dietrich, S.; Kempa, S.; Priewisch, B. *Sci. Synth.* 2007, *31b*, 1425-1537.

[289] Li, Z.; Wang, Q.; Tao, F. *Synth. Commun.* 2004, *34*, 1691-1702.

[290] Hegarty, A. F. *The Chemistry of Diazonium and Diazo Groups, Part 2*; John Wiley & Sons Ltd., 1978.

[291] Zollinger, H. *Azo and diazo chemistry: Aliphatic and aromatic compounds*; Interscience Publishers, 1961.

[292] Zollinger, P. D. H. *Diazo Chemistry I: Aromatic and Heteroaromatic Compounds*; Wiley-VCH, 1994.

[293] Butler, R. N. *Chem Rev.* 1975, *2*, 241-257.

[294] Shegal, I. L.; Stanovkina, K. V.; Kovalenko, N. G.; Shegal, L. M. *Chem. Heterocycl. Compd.* 1974, *10*, 369-371.

[295] Partridge, M. W.; Stevens, M. F. G. *J. Chem. Soc. C: Organic* 1967, 1828-1830.

[296] Huang, H.; Chen, F.; Chen, C.; Jiang, M. *Polymer Bulletin* 2016, *73*, 1545-1552.

[297] Wang, Y. Y.; Zhao, X. L.; Qian, H. F.; Huang. W. *Dyes Pigm.* 2018, *153*, 44-52.

[298] Přikryl, J.; Macháček, V.; Jansa, P.; Svobodová, M.; Růžička, A.; Nachtigall, P.; Čern, M. *Eur. J. Org. Chem.* 2008, 3272-3278.

[299] Raposo, M. M. M.; Ferreira A. M. F. P.; Amaro, M.; Belsley, M.;. Moura, J. C. V. P. *Dyes Pigm.* 2009, *83*, 59-65.

[300] Deligöz. H.; Ercan, N. *Tetrahedron* 2002, *58*, 2881-2884.

[301] Jina, C. M.; Lu, G. Y.; Wanga, Z. S.; Youa, X. Z.; Wangb, Z. H.; Gongb, W.; Wu, H. M. *Synthesis* 2001, *7*, 1023-1026.

[302] Şener, I.; Karcı, F.; Kılıç, E.; Deligöz, H. *Dyes Pigm.* 2004, *62*, 141-148.

[303] Yamada, S. *Formation of antisoiling acrylic silicone coatings on eyeglass frames,* JP 2007298663, 2007.

[304] 304) Lin, C.; Yang, L.; Xu, M.; An, Q.; Xiang, Z.; Liu, X. *RSC Adv.* 2016, *6*, 51552-51561.

[305] 305) Verma, R. K.; Aggarwal, M.; Bansal, M.; Kaur, I. P. *Synth. Commun.* 2007, *37*, 1977-1982.

[306] 306) Paul, H.; Mukherjee, T.; Mukherjee, M.; Mondal, T. K.; Moirangthem, A.; Basu, A.; Zangrando, E.; Chattopadhyay, P. *J. Coord. Chem.* 2013, *66*, 2747-2764.

[307] Stýskala, J.; Slouka, J.; Lycka, A. *J. Heterocycl. Chem.* 2009, *39*, 1305-1308.

[308] Huang, L.; Rominger, F.; Rudolph, M.; Hashmi, A. S. K. *Chem. Commun.* 2016, *52*, 6435-6438.

[309] Mahesh, V.K. *J. Indian Chem.Soc.* 1982, *59*, 1192-1193.

[310] Saylam, A.; Seferoğlu, Z.; Ertan, N. *J. Mol. Liq.* 2014, *195*, 267-276.

[311] Seferoğlu, Z.; Yalçın, E.; Babür, B.; Seferoğlu, N.; Hökelek, T.; Yılmaz, E.; Şahin, E. *Spectrochimica Acta Part A* 2013, *113*, 314-324.

[312] Li, W.; Bellera, M.; Wu, X. F. *Chem. Commun.* 2014, *50*, 9513-9516.

[313] Lazny, R.; Poplawski, J.; Köbberling, J.; Enders, D.; Bräse, S. *Synlet*t 1999, *8*, 1304-1306.

[314] Sarkar, D.; Pramanik, A. K.; Monda, T K. *Spectrochima Acta Part A* 2016, 153, 397-401.

[315] Zollinger, I. S. a. H. *Top. Curr. Chem* 1983, *112*.

[316] Liu, C.; Lv, J.; Luo, S.; Cheng, J.-P. *Org. Lett.* 2014, *16*, 5458-5461.

[317] Tskhovrebov, A. G.; Naested, L. C. E.; Solari, E.; Scopelliti, R.; Severin, K. *Angew. Chem. Int. Ed.* 2014, *54*, 1289-1292.

[318] Eymann, L. Y. M.; Scopelliti, R.; Tirani, F. F.; Severin, K. *Chem. Eur. J.* 2018, *24*, 7957-7963.

[319] Shchegol'kov, E. V.; Burgart, Y. V.; Khudina, O. G.; Saloutin, V. I.; Chupakhin, O. N. *Russ. Chem. Bull.* 2004, *53*, 2584-2590.

[320] Sayed, A. Z.; Aboul-Fetouh, M. S.; Nassar, H. S. *J. Mol. Struct.* 2012, *1010*, 146-151.

[321] Makhseeda, S.; Hassaneen, H. M. E.; Elnagdib, M. H. Z. *Naturforsch. B Chem. Sci.* 2007, *62*, 529-536.

[322] Semichenko, E. S.; Root, E. V.; Pokrovskii, L. M.; Suboch, G. A. *Russ. J. Org. Chem.* 2007, *43*, 406-408.

[323] Fokin, A. S.; Burgart, Y. V.; Saloutin, V. I. *Russ. J. Org. Chem.* 2006, *42*, 887-896.

[324] Zahedifar, M.; Razavi, R.; Sheibani, H. *J. Mol. Struct.* 2016, *1125*, 730-735.

[325] Şener, N.; Mohammed, H. J. A.; Yerlikaya, S.; Celik Altunoglu, Y.; Gür, M.; Baloglu, M. C.; Şener, I. *Dyes Pigm.* 2018, *157*, 11-19.

[326] Li, H.-H.; Bai, F.; Yan, H.; Lu, C.-S. *Org. Lett.* 2017, *19*, 862-865.

[327] Garst, M. E.; Lukton, D. *Synth. Commun.* 1980, *10*, 155-160.

[328] Cohen, T.; Lewarchik, R. J.; Tarino, J. Z. *J. Am. Chem. Soc.* 1974, *96*, 7753-7760.

[329] Curtin, D. Y.; Ursprung, J. A. *J. Org. Chem.* 1956, *21*, 1221-1225.

[330] Hansen, M. J.; Lerch, M. M.; Szymanski, W.; Feringa, B. L. *Angew. Chem. Int. Ed.* 2016, *55*, 13514-13518.

[331] Fischer, D.; Klapötke, T. M.; Stierstorfer, J. *Angew. Chem. Int. Ed.* 2015, *54*, 10299-10302.

[332] Faustino, H.; Brannigan, C. R.; Reis, L. V.; Santos, P. F.; Almeida, P. *Dyes Pigm.* 2009, *83*, 88-94.

[333] Faustino, H.; El-Shishtawy, R. M.; Reis, L. V.; Santos, P. F.; Almeida, P. *Tetrahedron Lett.* 2008, *49*, 6907-6909.

[334] Ueno, K.; Akiyoshi, S. *J. Am. Chem. Soc.* 1954, *76*, 3670–3672.

[335] Lu, T.-H.; Misra, T. K.; Lin, P.-C.; Liao, F.-L.; Chung, C.-S. *Polyhedron* 2003, *22*, 535-541.

[336] Roy, R.; Santra, P. K.; Das, D.; Sinha, C.; Sakata, K.; Kuma, H. *Synthesis and Reactivity in Inorganic and Metal-Organic Chemistry* 2000, *30*, 1975-1993.

[337] Obijalska, E.; Kaszynski, P.; Jankowiak, A.; Young, V. G. *Polyhedron* 2011, *30*, 1339-1348.

[338] Campbell, N.; Henderson, A. W.; Taylor, D. *J. Chem. Soc.* 1953, *0*, 1281-1285.

[339] Priewisch, B.; Rück-Braun, K. *J. Org. Chem.* 2005, *70*, 2350-2352.

[340] Yu, B.-C.; Shirai, Y.; Tour, J. M. *Tetrahedron* 2006, *62*, 10303-10310.

[341] J.Thiele, *Ber. Dtsch. Chem. Ges.* 1893, *26*, 2645-2647.

[342] Qu, Y.; Babailov, S. P. *J. Mat. Chem. A* 2018, *6*, 1915-1940.

[343] Hammerl, A.; Holl, G.; Klapötke, Thomas M.; Mayer, P.; Nöth, H.; Piotrowski, H.; Warchhold, M. *Eur. J. Inorg. Chem.* 2002, *2002*, 834-845.

[344] Klapötke, T. M.; Sabaté, C. M. *Chem. Mater.* 2008, *20*, 1750-1763.

[345] Williams, M. M.; McEwan, W. S.; Henry, R. A. *J. Chem. Phys.* 1957, *61*, 261-267.
[346] Krishna Mohan, V.; Tang, T. B. *J. Chem. Phys.* 1983, *79*, 4271-4278.
[347] Chaudhri, M. M. *Nature* 1976, *263*, 121.
[348] Chaudhri, M. M. *J. Mater. Sci.* 1984, *19*, 3028-3042.
[349] López-Tarifa, P.; Sánchez-Sanz, G.; Alkorta, I.; Elguero, J.; Sanz, D.; Perona, A.; Claramunt, R. M. *J. Mol. Struct.* 2012, *1015*, 138-146.
[350] Klapötke, T. M.; Piercey, D. G. *Inorg. Chem.* 2011, *50*, 2732-2734.
[351] Batog, L. V.; Konstantinova, L. S.; Rozhkov, V. Y. *Russ. Chem. Bull.* 2005, *54*, 1915-1922.
[352] Glover, E. E.; Rowbottom, K. T. *J. Chem. Soc., Perkin Trans. 1* 1974, 1792-1794.
[353] Loew, P.; Weis, C. D. *J. Heterocycl. Chem.* 1976, *13*, 829-833.
[354] Kniep, F.; Walter, S. M.; Herdtweck, E.; Huber, S. M. *Chem. Eur. J.* 2012, *18*, 1306-1310.
[355] Liu, G.-F.; Ji, W.; Feng, C.-L. *Langmuir* 2015, *31*, 7122-7128.
[356] Drug, E.; Gozin, M. *J. Am. Chem. Soc.* 2007, *129*, 13784-13785.
[357] Dutta, B.; Biswas, S.; Sharma, V.; Savage, N. O.; Alpay, S. P.; Suib, S. L. *Angew. Chem. Int. Ed.* 2016, *55*, 2171-2175.
[358] Gattavecchia, E.; Tonelli, D. *J. Chem. Soc., Perkin Trans. 2* 1986, 689-693.
[359] Weston, C. E.; Richardson, R. D.; Fuchter, M. J. *Chem. Commun.* 2016, *52*, 4521-4524.
[360] Pizzolatti, M. G.; Yunes, R. A. *J. Chem. Soc., Perkin Trans. 2*, 1990, 759-764.
[361] 361) Merino, E. *Chem. Soc. Rev.* 2011, *40*, 3835-3853.
[362] Huynh, M. H. V.; Hiskey, M. A.; Pollard, C. J.; Montoya, D. P.; Hartline, E. L.; Gilardi, R. *J. Energ Mater.* 2004, *22*, 217-229.
[363] Huynh, M.-H. V.; Hiskey, M. A.; Hartline, E. L.; Montoya, D. P.; Gilardi, R. *Angew. Chem. Int. Ed.* 2004, *43*, 4924-4928.
[364] Lim, Y.-K.; Choi, S.; Park, K. B.; Cho, C.-G. *J. Org. Chem.* 2004, *69*, 2603-2606.
[365] Garcia, J. A.; Lobo, M. D.; Nonell, S.; Velasco, D. *Angew. Chem. Int. Ed.* 2012, *51*, 12820-12823.

[366] Travieso-Puente, R.; Budzak, S.; Chen, J.; Stacko, P.; Jastrzebski, J. T. B. H.; Jacquemin, D.; Otten, E. *J. Am. Chem. Soc.* 2017, *139*, 3328-3331.

[367] Chang, M. C.; Otten, E. *Chem. Commun.* 2014, *50*, 7431-7433.

[368] Barak, D. S.; Dighe, S. U.; Avasthi, I.; Batra, S. *J. Org. Chem.* 2018, *83*, 3537-3546.

[369] Huang, P.-K. C.; Kosower, E. M. *J. Am. Chem. Soc.* 1967, *89*, 3910-3911.

[370] Huang, P.-K. C.; Kosower, E. M. *J. Am. Chem. Soc.* 1968, *90*, 2367-2376.

[371] Kosower, E. M.; Huang, P.-K. C.; Tsuji, T. *J. Am. Chem. Soc.* 1969, *91*, 2325-2329.

[372] Jasch, H.; Scheumann, J.; Heinrich, M. R. *J. Org. Chem.* 2012, *77*, 10699-10706.

[373] Hofmann, J.; Jasch, H.; Heinrich, M. R. *J. Org. Chem.* 2014, *79*, 2314-2320.

[374] Pandit, R. P.; Lee, Y. R. *Adv. Synth. Catal.* 2015, *357*, 2657-2664.

[375] Pandit, R. P.; Kim, S. H.; Lee, Y. R. *Org. Biomol. Chem.* 2016, *14*, 6996-7000.

[376] Powers, I. G.; Andjaba, J. M.; Luo, X.; Mei, J.; Uyeda, C. *J. Am. Chem. Soc.* 2018, *140*, 4110-4118.

[377] Al-Jalal, N. A.; Ibrahim, M. R.; Al-Awadi, N. A.; Elnagdi, M. H. *Molecules* 2011, *16*, 10256-10268.

[378] Yoshida, Z.; Hashimoto, H.; Yoneda, S. *J. Chem. Soc. D: Chem. Commun.* 1971, 1344-1345.

[379] Hünig, S.; Oette, K-H. *Justus Liebigs Ann. Chem.* 1961, *641*, 104-112.

[380] Hnig, S.; *Angew. Chem.* 1961, *73*, 818 – 824.

[381] Hünig, H.; Balli, H.; Fritsch, K. H.; Herrmann, H.; Köbrich, G.; Werner, H.; Grigat, E.; Müller, F.; Nothher, H.; Oette, K. H. *Angew. Chem.* 1958, *70*, 215-222.

[382] Siewertsen, R.; Neumann, H.; Buchheim-Stehn, B.; Herges, R.; Näther, C.; Renth, F.; Temps F. *J. Am. Chem. Soc.* 2009, *131*, 15594–15595.

[383] *Norikane, Y.; Katohb, R.; Tamaoki, N. Chem. Commun*. 2008, *0*, 1898-1900.

[384] 384) Bortolus, P.;. Monti, S. *J. Phys. Chem.* 1979, 83, 648–652.

[385] Rau, H. *J. Photochem.* 1984, *26*, 221-5.

[386] Lednev, I. K.; Ye, T.-Q.; Hester, R. E.; Moore, J. N. *J. Phys. Chem.* 1996, *100*, 13338-13341.

[387] Lednev, I. K.; Ye, T. Q.; Matousek, P.; Towrie, M.; Foggi, P.; Neuwahl, F. V. R.; Umapathy, S.; Hester, R. E.; Moore, J. N. *Chem. Phys. Lett.* 1998, *290*, 68-74.

[388] Fujino, T.; Arzhantsev, S. Y.; Tahara, T. *J. Phys. Chem. A* 2001, *105*, 8123-8129.

[389] Fujino, T.; Arzhantsev, S. Y.; Tahara, T. *Bull. Chem. Soc. Jpn.* 2002, *75*, 1031-1040.

[390] Fujino, T.; Tahara, T. *J. Phys. Chem. A* 2000, *104*, 4203-4210.

[391] Schultz, T.; Quenneville, J.; Levine, B.; Toniolo, A.; Martinez, T. J.; Lochbrunner, S.; Schmitt, M.; Shaffer, J. P.; Zgierski, M. Z.; Stolow, A. *J. Am. Chem. Soc.* 2003, *125*, 8098-8099.

[392] Stuart, C. M.; Frontiera, R. R.; Mathies, R. A. *J. Phys. Chem. A* 2007, *111*, 12072-12080.

[393] Chang, C.-W.; Lu, Y.-C.; Wang, T.-T.; Diau, E. W.-G. *J. Am. Chem. Soc.* 2004, *126*, 10109-10118.

[394] Harabuchi, Y.; Ishii, M.; Nakayama, A.; Noro, T.; Taketsugu, T. *J. Chem. Phys.* 2013, *138*, 064305.

[395] Crecca, C. R.; Roitberg, A. E. *J.Phys. Chem. A* 2006, *110*, 8188-8203.

[396] Tan, E. M. M.; Amirjalayer, S.; Smolarek, S.; Vdovin, A.; Buma, W. J.; Zerbetto, F. *Nat. Commun.* 2015, *6*, 5860.

[397] Nenov, A.; Borrego-Varillas, R.; Oriana, A.; Ganzer, L.; Segatta, F.; Conti, I.; Segarra-Marti, J.; Omachi, J.; Dapor, M.; Taioli, S.; Manzoni, C.; Mukamel, S.; Cerullo, G.; Garavelli, M. *J. Phys. Chem. Lett.* 2018, *9*, 1534-1541.

[398] Gámez, J. A.; Weingart, O.; Koslowski, A.; Thiel, W. *J. Chem. Theory Comput.* 2012, *8*, 2352-2358.

[399] Ishikawa, T.; Noro, T.; Shoda, T. *J. Chem. Phys.* 2001, *115*, 7503-7512.

[400] Wang, Y.-T.; Liu, X.-Y.; Cui, G.; Fang, W.-H.; Thiel, W. *Angew. Chem. Int. Ed.* 2016, *55*, 14009-14013.
[401] Wei-Guang Diau, E. *J. Phys. Chem. A* 2004, *108*, 950-956.
[402] Monti, S.; Orlandi, G.; Palmieri, P. *Chem. Phys.* 1982, *71*, 87-99.
[403] Schulte-Frohlinde, D. *Justus Liebigs Ann. Chem.* 1958, *612*, 131-138.
[404] Talaty, E. R.; Fargo, J. C. *Chem. Commun.* 1967, 65-66.
[405] Whitten, D. G.; Wildes, P. D.; Pacifici, J. G.; Irick, G., Jr. *J. Am. Chem. Soc.* 1971, *93*, 2004-2008.
[406] Asano, T.; Okada, T.; Shinkai, S.; Shigematsu, K.; Kusano, Y.; Manabe, O. *J. Am. Chem. Soc.* 1981, *103*, 5161-5165.
[407] Haberfield, P.; Block, P. M.; Lux, M. S. *J. Am. Chem. Soc.* 1975, *97*, 5804-5806.
[408] Nishimura, N.; Sueyoshi, T.; Yamanaka, H.; Imai, E.; Yamamoto, S.; Hasegawa, S. *Bull. Chem. Soc. Jpn.* 1976, *49*, 1381-1387.
[409] Joshi, N. K.; Fuyuki, M.; Wada, A. *J. Phys. Chem. B* 2014, *118*, 1891-1899.
[410] Otsuki, J.; Suwa, K.; Narutaki, K.; Sinha, C.; Yoshikawa, I.; Araki, K. *J. Phys. Chem. A* 2005, *109*, 8064-8069.
[411] Otsuki, J.; Suwa, K.; Sarkar, K. K.; Sinha, C. *J. Phys. Chem. A* 2007, *111*, 1403-1409.
[412] Čechová, L.; Kind, J.; Dračínský, M.; Filo, J.; Janeba, Z.; Thiele, C. M.; Cigáň, M.; Procházková, E. *J. Org. Chem.* 2018, *83*, 5986-5998.
[413] Yu, X.; Wang, Z.; Buchholz, M.; Füllgrabe, N.; Grosjean, S.; Bebensee, F.; Bräse, S.; Wöll, C.; Heinke, L. *Phys. Chem. Chem. Phys.* 2015, *17*, 22721-22725.
[414] Quick, M.; Dobryakov, A. L.; Gerecke, M.; Richter, C.; Berndt, F.; Ioffe, I. N.; Granovsky, A. A.; Mahrwald, R.; Ernsting, N. P.; Kovalenko, S. A. *J. Phys. Chem. B* 2014, *118*, 8756-8771.
[415] Ebadi, H. *J. Phys. Chem. A* 2014, *118*, 7832-7837.
[416] Corchado, J. C.; Sanchez, M. L.; Fdez. Galvan, I.; Martin, M. E.; Munoz-Losa, A.; Barata-Morgado, R.; Aguilar, M. A. *J. Phys. Chem. B* 2014, *118*, 12518-12530.
[417] Ikegami, T.; Kurita, N.; Sekino, H.; Ishikawa, Y. *J. Phys. Chem. A* 2003, *107*, 4555-4562.

[418] Bull, J. N.; Scholz, M. S.; Coughlana, N. J. A.; Bieske, E.J. *Phys. Chem. Chem. Phys*. 2017, *19*, 12776-12783.

[419] Pang, X.; Jiang, C.; Qi, Y.; Yuan, L.; Hu, D.; Zhang, X.; Zhao, D.; Wang, D.; Lan, Z.; Lia, F. *Phys. Chem. Chem. Phys*. 2018, *20*, 25910-25917.

[420] Zhou, L.; Liu, J.; Zhou, P. *J. Phy. Chem. A* 2017, *121*, 141-150.

[421] Garcia-Amorós, J.; Nonell, S.; Velasco, D. *Chem. Commun*. 2012, *48*, 3421-3423.

[422] Ryan, A. *Br. J. Pharmacol*. 2017, *174*, 2161-2173.

[423] Sharma, R.; Rawal, R. K.; Gaba, T.; Singla, N.; Malhotra, M.; Matharoo, S.; Bhardwaj, T. R. *Bioorg. Med. Chem. Lett*. 2013, *23*, 5332-5338.

[424] Velema, W. A.; van der Berg, J. P.; Hansen, M. J.; Szymanski, W.; Driessen, A. J. M.; Feringa, B. L. *Nat. Chem*. 2013, *5*, 924-928.

[425] Weston, C. E.; Kramer, A.; Colin, F.; Yildiz, O.; Baud, M. G. J.; Meyer-Almes, F.-J.; Fuchter, M. J. *ACS Infect. Dis*. 2017, *3*, 152-161.

[426] Zeng, L.; Kuang, S.; Li, G.; Jin, C.; Ji, L.; Chao, H. *Chem. Commun*. 2017, *53*, 1977-1980.

[427] Romero-Canelón, I.; Sadler, P. J. *Inorg. Chem*. 2013, *52*, 12276-12291.

[428] Fu, Y.; Romero, M. J.; Habtemariam, A.; Snowden, M. E.; Song, L.; Clarkson, G. J.; Qamar, B.; Pizarro, A. M.; Unwin, P. R.; Sadler, P. J. *Chem. Sci*. 2012, *3*, 2485-2494.

[429] Matera, C.; Gomila, A. M. J.; Camarero, N.; Libergoli, M.; Soler, C.; Gorostiza, P. *J. Am. Chem. Soc*. 2018, *140*, 15764–15773.

[430] Engel, S.; Möller, N.; Stricker, L.; Peterlechner, M.; Ravoo, B. J. *Small* 2018, *14*, 1704287.

[431] Stricker, L.; Fritz, E.-C.; Peterlechner, M.; Doltsinis, N. L.; Ravoo, B. J. *J. Am. Chem. Soc*. 2016, *138*, 4547-4554.

[432] Wiemann, M.; Niebuhr, R.; Juan, A.; Cavatorta, E.; Ravoo, B. J.; Jonkheijm, P. *Chem. Eur. J*. 2018, *24*, 813-817.

[433] Moeller, N.; Hellwig, T.; Stricker, L.; Engel, S.; Fallnich, C.; Ravoo, B. J. *Chem. Commun*. 2017, *53*, 240-243.

[434] Adam, V.; Prusty, D. K.; Centola, M.; Skugor, M.; Hannam, J. S.; Valero, J.; Kloeckner, B.; Famulok, M. *Chem. Eur. J.* 2018, *24*, 1062-1066.

[435] Hutchins, K. M.; Kummer, K. A.; Groeneman, R. H.; Reinheimer, E. W.; Sinnwell, M. A.; Swenson, D. C.; MacGillivray, L. R. *Cryst Eng Comm* 2016, *18*, 8354-8357.

[436] Pfletscher, M.; Woelper, C.; Gutmann, J. S.; Mezger, M.; Giese, M. *Chem. Commun.* 2016, *52*, 8549-8552.

[437] Chen, Y.; Yu, H.; Zhang, L.; Yang, H.; Lu, Y. *Chem. Commun.* 2014, *50*, 9647-9649.

[438] Du, M.; Li, L.; Zhang, J.; Li, K.; Cao, M.; Mo, L.; Hu, G.; Chen, Y.; Yu, H.; Yang, H. *Liq. Cryst.* 2018, 1-8.

[439] Shinkai, S.; Nakaji, T.; Ogawa, T.; Shigematsu, K.; Manabe, O. *J. Am. Chem. Soc.* 1981, *103*, 111-115.

[440] Chang, Z.; Zhang, D.-S.; Chen, Q.; Li, R.-F.; Hu, T.-L.; Bu, X.-H. *Inorg. Chem.* 2011, *50*, 7555-7562.

[441] Mohamed, M. H.; Elsaidi, S. K.; Pham, T.; Forrest, K. A.; Schaef, H. T.; Hogan, A.; Wojtas, L.; Xu, W.; Space, B.; Zaworotko, M. J.; Thallapally, P. K. *Angew. Chem. Int. Ed.* 2016, *55*, 8285-8289.

[442] He, C.; Zhang, B.-G.; Duan, C.-Y.; Li, J.-H.; Meng, Q.-J. *Eur. J. Inorg. Chem.* 2000, 2549-2554.

[443] Suzuki, R.; Onodera, T.; Kasai, H.; Oikawa, H. *Mol. Cryst. Liq. Cryst.* 2017, *654*, 109-114.

[444] Mohamed, M. G.; Su, W.-C.; Lin, Y.-C.; Wang, C.-F.; Chen, J.-K.; Jeong, K.-U.; Kuo, S.-W. *RSC Adv.* 2014, *4*, 50373-50385.

[445] Peters, M. V.; Stoll, R. S.; Kuehn, A.; Hecht, S. *Angew. Chem. Int. Ed.* 2008, *47*, 5968-5972.

[446] Stoll, R. S.; Peters, M. V.; Kuhn, A.; Heiles, S.; Goddard, R.; Buhl, M.; Thiele, C. M.; Hecht, S. *J. Am. Chem. Soc.* 2009, *131*, 357-367.

[447] Dommaschk, M.; Gutzeit, F.; Boretius, S.; Haag, R.; Herges, R. *Chem. Commun.* 2014, *50*, 12476-12478.

[448] Dommaschk, M.; Thoms, V.; Schütt, C.; Näther, C.; Puttreddy, R.; Rissanen, K.; Herges, R. *Inorg. Chem.* 2015, *54*, 9390-9392.

[449] Thies, S.; Sell, H.; Schütt, C.; Bornholdt, C.; Näther, C.; Tuczek, F.; Herges, R. *J. Am. Chem. Soc.* 2011, *133*, 16243-16250.

[450] Ragon, F.; Yaksi, K.; Sciortino, N. F.; Chastanet, G.; Létard, J. -F.; D'Alessandro, D. M.; Kepert, C. J.; Neville, S. M. *Aust. J. Chem.* 2014, *67*, 1563-1573.

[451] Heitmann, G.; Schütt, C.; Gröbner, J.; Huber, L.; Herges, R. *Dalton Trans.* 2016, *45*, 11407-11412.

[452] Hasegawa, Y.; Kume, S.; Nishihara, H. *Dalton Trans.* 2009, 280-284.

[453] Bello, K. A.; Griffiths, J. *J. Chem. Soc., Chem. Commun.* 1986, 1639-1640.

[454] Li, Y.; Patrick, B. O.; Dolphin, D. *J. Org. Chem.* 2009, *74*, 5237-5243.

[455] Salvador, M. A.; Almeida, P.; Reis, L. V.; Santos, P. F. *Dyes Pigm.* 2009, *82*, 118-123.

[456] Dong, M.; Babalhavaeji, A.; Samanta, S.; Beharry, A. A.; Woolley, G. A. *Acc. Chem. Res.* 2015, *48*, 2662-2670.

[457] Yang, Y.; Hughes, R. P.; Aprahamian, I. *J. Am. Chem. Soc.* 2012, *134*, 15221-15224.

[458] Yang, Y.; Hughes, R. P.; Aprahamian, I. *J. Am. Chem. Soc.* 2014, *136*, 13190-13193.

[459] Cai, J.; Li, Z.; Qiu, Y.; OuYang, Z.; Lin, W.; Yang, L.; Feng, W.; Yu, X.; Dong, W. *New J. Chem.* 2016, *40*, 9370-9379.

[460] Rauf, M. A.; Hisaindee, S.; Saleh, N. *RSC Adv.* 2015, *5*, 18097-18110.

[461] Dong, M.; Babalhavaeji, A.; Collins, C. V.; Jarrah, K.; Sadovski, O.; Dai, Q.; Woolley, G. A. *J. Am. Chem. Soc.* 2017, *139*, 13483–13486.

[462] Devi, S.; Gaur, A. K.; Gupta, D.; Saraswat, M.; Venkataramani, S. *Chem Photo Chem* 2018, *2*, 806-810.

[463] Naskar, K.; Maity, S.; Jana, S.; Dutta, B.; Tanaka, S.; Mallick, D.; Akitsu, T.; Sinha, C. *Cryst. Growth Des.* 2018, *18*, 2986-2997.

[464] Suzuki, T.; Moriya, T.; Endo, R.; Iwasaki, N. *Polym. Chem.* 2017, *8*, 761-768.

[465] Horváth, M.; Cigáň, M.; Filo, J.; Jakusová, K.; Gáplovský, M.; Šándrik, R.; Gáplovský, A. *RSC Adv.* 2016, *6*, 109742-109750.

[466] Bartwal, G.; Aggarwal, K.; Khurana, J. M. *New J. Chem.* 2018, *42*, 2224-2231.

[467] E. Rudenko, A.; Clayman, N.; Maclaren, J.; Waymouth, R. *Reversible Electropolymerization of Nickel Complexes Based on Redox-Mediated Ligand Exchange*, 2016, 1.

[468] Giridhar, M.; Naik, H. S. B.; Sudhamani, C. N.; Prabhakar, M. C.; Kenchappa, R.; Patil, S. *J. Chem. Pharm. Res.* 2017, *9*, 137-142.

In: Photoisomerization
Editor: Diego Sampedro

ISBN: 978-1-53615-313-2

Chapter 5

RETINAL INSPIRED PHOTOSWITCHES: FROM THE ISOMERIZATION MECHANISMS TOWARD RECENT APPLICATIONS

***Marco Marazzi*[1,2] *and Cristina García-Iriepa*[3,*]**
[1]Departamento de Química, Centro de Investigación en Síntesis Química (CISQ), Universidad de La Rioja, Logroño, Spain
[2]Departamento de Química Analítica, Química Física e Ingeniería Química, Universidad de Alcalá, Alcalá de Henares, Spain
[3]Laboratoire Modélisation et Simulation Multi Echelle, Université Paris-Est, MSME, UPEM, Marne-la-Vallée, France

ABSTRACT

Retinal is the protonated Schiff base known to be the covalently bound chromophore of rhodopsins, a class of transmembrane proteins involved in several animal and microbial light-induced processes. Depending on the type of rhodopsin, different retinal isomers constitute the core of the protein opsin pocket. In all cases, a light-induced *E*/*Z* photoisomerization

[*] Corresponding Author's Email: marco.marazzi@uah.es; cristina.garciairiepa@u-pem.fr.

is responsible for the activation of the biological signal, involved in visual phototransduction, phototrophy and, more recently, optogenetic applications. In this reviewing chapter, we will describe the mechanisms underlying retinal isomerization in their natural rhodopsin environment, as a source of inspiration to build biomimetic photoswitches. Especially, their application as single molecule in the biological field will be described, focusing on the capacity to control the secondary structure of selected peptides. Moreover, depending on the environment, retinal based photoswitches can also be proposed as building blocks for new generations of highly efficient molecular motors.

1. INTRODUCTION

The *E/Z* photoisomerization of a specific double bond is a quite common process in nature (Dugave 2006; Dugave and Demange 2003). Some examples of light-sensitive proteins are rhodopsins (involved, among others, in the process of vision) (Schoenlein et al. 1991; Ernst et al. 2014), the photoactive yellow protein (blue-light receptor) (Heyne et al. 2005; Kort et al. 1996), the green fluorescent protein (where the photoisomerization is a competitive pathway to light emission) (Rafiq et al. 2011; Voliani et al. 2008) and phytochromes (light receptors in plants, bacteria and fungi) (Quail et al. 1995; Rohmer et al. 2010). In order to be light-sensitive, these proteins host a chromophore which, upon light absorption, undergoes an *E/Z* photoisomerization. This photochemical reaction causes a large change of the chromophore's structure leading to a local or overall conformational change of the light-sensitive protein. In most cases, this change constitutes the first event of a cascade of biochemical reactions.

Rhodopsin is one of the most studied photoreceptor proteins due to its high selectivity and efficiency (Gozem et al. 2017). In particular, light is absorbed by the retinal chromophore, a protonated Schiff base located within rhodopsin (Figure 1). Upon light absorption, retinal undergoes a quite efficient and ultrafast *E/Z* photoisomerization around one of its carbon-carbon double bonds. Moreover, it has been shown that the protein surrounding plays an active role in modulating photophysical and photochemical properties of this reaction (Melaccio et al. 2016). For all these reasons, the photoisomerization process of the retinal chromophore within rhodopsin has been selected for numerous studies, both experimental and theoretical, as we will describe in the following section. In this chapter, we will go through the details of the rhodopsins' response to light, giving an overview of the different rhodopsin types, including a description of the various ground state hydrogen bonding patterns, as well as the conceived photoisomerization models.

The highly efficient photoisomerization of the retinal chromophore has provided a solid source of inspiration for the design of photoactive molecular switches. The study of molecular machines, and among them of molecular switches, has attracted the interest of numerous scientists covering the design, the study of fundamental properties and the development of smart applications (Browne and Feringa 2006; Balzani, Credi, and Venturi 2009). The relevance of this field was demonstrated by the Nobel Prize in Chemistry 2016, awarded to Jean-Pierre Sauvage, Sir J. Fraser Stoddart and Bernard L. Feringa for the design and synthesis of molecular machines. In general terms, a molecular switch is a molecule that can be reversibly interconverted between two states when an energy input is applied (Figure 2A). When light is used as the input, these molecules are also known as photoswitches. Actually, the retinal chromophore could be considered as a natural photoswitch as it can be interconverted between the *E* and *Z* isomers by light absorption. For this reason, molecular switches that mimic the protonated Schiff base of the retinal chromophore are known as retinal-like photoswitches, since they have been designed as non-natural derivatives aimed to replicate the properties and photochemical processes found *in vivo* (García-Iriepa, Marazzi, Frutos, et al. 2013). In this chapter, the different families of molecular switches based on the retinal chromophore will be presented. In particular, we will focus our attention on the synthetic roots and on their photophysical and photochemical properties.

Thanks to the variety of retinal-like photoswitches and the corresponding wide range of properties, diverse applications have been proposed. It outstands the use of this type of photoswitches to photocontrol the secondary structure of a given peptide. Indeed, it has been shown that the photoisomerization of a retinal-like switch covalently linked to a peptide induces a significant change in its structure, with the concomitant modification of the peptide conformation (Figure 2B) (Blanco-Lomas et al. 2012; García-Iriepa et al. 2016; Andruniów et al. 2005; Adalgisa Sinicropi et al. 2009). This conformational change of the peptide could modify its biological activity. In this chapter we describe the examples reported until now regarding the use of retinal-like photoswitches to photocontrol the secondary structure of selected peptides.

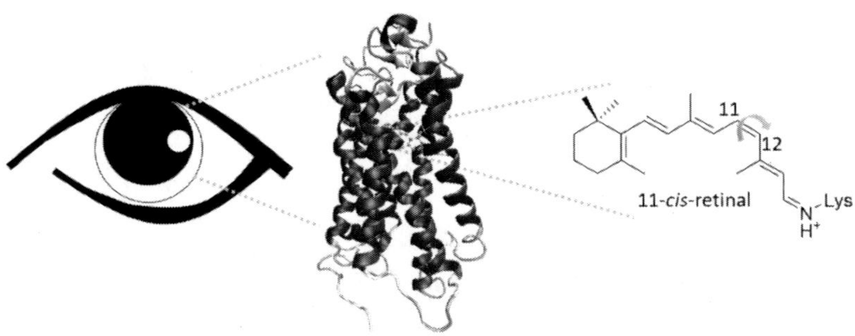

Figure 1. Graphical representation of the rhodopsin protein involved in the process of vision (bovine rhodopsin, PDB code: 1U19). Within the protein, the retinal chromophore keeps an initial 11-*cis* conformation prior to photon absorption. The photoisomerizable double bond is depicted with an arrow, leading toward the all-*trans* final conformation.

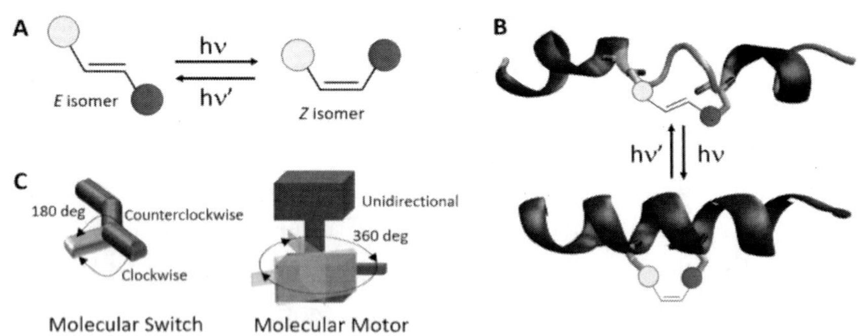

Figure 2. A) Schematic representation of an *E/Z* molecular photoswitch. B) Peptide conformational change upon switch photoisomerization. C) Comparison of the non-controlled (clockwise or counterclockwise) and controlled (unidirectional) movement of molecular switches and motors, respectively.

Finally, the natural photoisomerization reaction of retinal inside rhodopsins has served as source of inspiration to design other types of molecular devices: molecular motors. The main difference between a molecular switch and a motor concerns its mechanism. In particular, within photoswitches the interconversion between the two states implies a non-controlled (clockwise and counterclockwise directions) rotation of 180 degrees. This happens especially in homogeneous isotropic environments such as in gas phase or in solution. In the case of anisotropic environments, as the retinal embedded in its rhodopsin pocket, the photoswitch will have a preferred rotation direction. On the other hand, a molecular motor is characterized by a controlled, either clockwise or counterclockwise, rotation of 360 degrees. Based on this idea, different approaches have been proposed to convert the non-controlled rotation of a retinal-like switch into a controlled one, leading to a molecular motor. In this chapter we will present the strategies devised until now for the design of retinal-like motors and their properties (Kazaryan et al. 2010; Graydon 2015; Sluysmans and Stoddart 2018; Erbas-Cakmak et al. 2015).

2. Retinal Isomerization in Rhodopsins

Some proteins as rhodopsins were optimally designed by Nature in order to control their activity by light. Rhodopsins are transmembrane proteins which monomeric form is composed by seven α-helices. Especially, the retinal is the photoisomerizable chromophore constituting the cofactor of all types of rhodopsins. In this sense, the retinal chromophore is a typical two-states molecular photoswitch, to which light can be applied in order to transform *state 1* (*cis* or *trans* initial conformation) into *state 2* (*trans* or *cis* final conformation). Thermal activation (dark reaction) can reverse the system back to the initial form (Andruniow, Ferre, and Olivucci 2004; Frutos et al. 2007; Schapiro et al. 2011; Nakamichi, Buss, and Okada 2007; Smitienko et al. 2010; Kandori, Shichida, and Yoshizawa 2001). In the biological field, reversibility of a photoswitch is of particular interest for control purposes: depending on the irradiation wavelength, it is possible to either activate or inhibit a certain interaction or function. An even deeper control could arise if the *state 2*-to-*state 1* dark reaction could be modified to be possibly activated with light of a different wavelength compared to the irradiation of *state 1*, at the same time inhibiting the thermal back reaction by increasing its activation energy (Blanco-Lomas et al. 2012; Sadovski et al. 2009; Schierling et al. 2010).

The retinal is covalently bound to a lysine residue via a protonated Schiff base linkage and its spectroscopic and photochemical properties depend largely on the amino acid side chains forming the opsin pocket around the chromophore. After photoisomerization, a proton transfer to the nearest counterion (a negatively charged amino acid group or a water molecule) initiates the biological photocycle leading to the accomplishment of the specific rhodopsin function. As stated in the introduction, vision is one of the most known biological functions that rhodopsins are responsible of. More in detail, visual rhodopsin is a G protein-coupled receptor responsible for vision in mammalians through an 11-*cis* to all-*trans* retinal photoisomerization.

Apart from visual rhodopsins of the animal kingdom, other types of rhodopsins are found in prokaryotes and algae, usually containing an all-*trans* retinal isomer which isomerizes to 13-*cis* when activated by light. Such microbial rhodopsins can be classified depending on their function: sensors as sensory rhodopsin II, responsible for generation of the phototaxis signal, i.e., movement in response to light (Sasaki and Spudich 2008); light-driven ion pumps as bacteriorhodopsins (bRs) (Matsuno-Yagi and Mukohata 1977), halorhodopsins (Oesterhelt and Stoeckenius 1971), proteorhodopsins (Slamovits et al. 2011) and xanthorhodopsins (Balashov and Lanyi 2007); direct light-driven ion-channels as channelrhodopsins (ChRs) (Nagel et al. 2002). In particular, sensors and light-driven ion pumps have been extensively studied, including structure, mechanism and function (Lanyi 2004; Neutze et al. 2002; Klare, Chizhov, and Engelhard 2008). On the other hand, a lot of questions about ChRs are still opened. This is mainly due to the difficulties encountered to obtain accurate ChR models. In fact, only two X-ray structures are available: a C1C2 hybrid (5 α-helices from ChR1 and 2 α-helices from ChR2, hence with structural properties more similar to

ChR1) (Kato et al. 2012) and the wild-type ChR2 (Volkov et al. 2017). ChR1 and ChR2 were both discovered in microalgae *Chlamydomonas Reinhardtii*, where they function as sensory photoreceptors by controlling phototaxis (Sineshchekov, Jung, and Spudich 2002). Their main differences relies on the easier expression of ChR2 in host organisms (Berthold et al. 2008) and on the pH-independent absorption properties of ChR2 (pH-range 4-9) (Ritter et al. 2008). Hence, ChR2 results in the most studied and applied type of channelrhodopsin for biophysical and optogenetics purposes. In any case, the structural arrangement of the retinal chromophore within the opsin is still under debate, even though important advances have been recently made (Kato et al. 2012; Watanabe et al. 2012; Welke et al. 2011; Guo et al. 2016). In the last decade, ChRs were largely known as the photoswitches that allowed the discovery of optogenetics, by which light pulses are used to control neuronal activity in a considerable range of animals, opening a wide scenario for understanding fundamental aspects of brain functions, dysfunctions and related pathologies. Indeed, when expressed in cells of other organisms, light can be applied to control calcium influx, transmembrane potential and intracellular acidity, among other photoelectrochemical processes (Fenno, Yizhar, and Deisseroth 2010; Hegemann and Möglich 2011; Yizhar et al. 2011).

Concerning retinal photoisomerization, various mechanistic scenarios are possible, all of them sharing a common denominator: a motion in the opsin cavity pointing toward volume conservation, being assured by the rotation of selected retinal carbon-carbon bonds. More in detail, Warshel et al. proposed a space saving mechanism in visual rhodopsin called "aborted bicycle pedal": the retinal starts a simultaneous and concerted rotation of two double bonds in opposite directions, but only one is completed (determining the photoisomerizable double bond) while the other one is aborted (Arieh Warshel 1976; Arieh Warshel and Barboy 1982; A. Warshel, Chu, and Hwang 1991). More recently, the "asynchronous bicycle pedal" photoisomerization mechanism was proposed by Frutos et al. (Frutos et al. 2007), where the two double bond rotations are not concerted. Also the "hula-twist" mechanism was proposed by Liu et al. (R. S. Liu and Asato 1985; R. S. H. Liu 2001), including translocation of a single C-H unit along the polyene chain, followed by the simultaneous rotation of the two connected bonds, one single and one double bond. Apart from the visual pigment, the isomerization of the sensor *Anabaena* sensory rhodopsin (ASR) was studied by Strambi et al. (Strambi et al. 2010) and Schapiro et al. (Schapiro and Ruhman 2014), reporting in both cases a unidirectional rotation of retinal. Since ASR has the peculiarity to bind both all-*trans* and 13-*cis* retinal isomers, the excited state kinetics was probed experimentally, leading to a slower photo-reaction for the 13-*cis* isomer (Agathangelou et al. 2018). Regarding ion pumps, the "double bicycle pedal" mechanism was proposed in bR by Altoé et al. (Altoe et al. 2010): the three adjacent double bonds C11 = C12, C13 = C14 and C15 = N rotate concomitantly in alternate directions. Twisting around C13 = C14 is compensated by the sum of C11 = C12 and C15 = N rotations, both in the opposite direction.

Regarding ion-channels, studies focused on ChR isomerization are emerging in the last years. Experimentally, the photochemistry of the hybrid C1C2 and of *Ps*ACR1 (the anion ChR) were studied by ultrafast spectroscopy (Hontani, Marazzi, et al. 2017; Hontani, Broser, et al. 2017). In the case of C1C2, the mechanism was explained also by computational techniques (Hontani, Marazzi, et al. 2017; Dokukina et al. 2018), leading to a different scenario with respect to previously described rhodopsins: the ChR ground state is characterized by structural heterogeneity of the retinal active site, as suggested by the computational finding of several ground state minima associated to a plethora of excited-state time constants for the decay to the ground state. The flexible active site structure can be rationalized by different hydrogen bonding patterns around the protonated Schiff base, in contrast to the single pattern shown by the rest of rhodopsins. For example, bR is well known for its stable three-water pentagonal active site, as shown computationally and by experiments (Figure 3E) (Shibata, Tanimoto, and Kandori 2003). On the other hand, in ChR2 three different patterns were proposed by extensive molecular dynamics (MD), both classical MD and more accurate hybrid quantum mechanics/molecular mechanics (QM/MM) MD (Guo et al. 2016): a direct hydrogen bond between the protonated Schiff base ($-NH^+$) and (*i*) a glutamic acid (Figure 3B), (*ii*) an aspartic acid (Figure 3C) or (*iii*) a water molecule (Figure 3D). This is a critical issue, since the counterion of the protonated Schiff base largely affects the spectroscopic and possibly isomerization properties.

As a matter of fact, various models were proposed to explain the nature of the excited states involved in the photoisomerization process. The main difference is the number of electronic states required. At least two states are needed, as in the visual pigment: the ground state (S_0) and the lowest singlet excited state (S_1), of charge transfer character. Indeed, the positive charge located on the Schiff base tends to migrate along the polyene chain, toward the β-ionone ring (see Figure 1). After $S_0 \rightarrow S_1$ absorption the system evolves on S_1 undergoing bond length alternation of the polyene chain, i.e., single bonds become double and double bonds become single. This first step, usually assigned to a characteristic spectroscopic time of some tents of femtoseconds, can be complete or only partial. In the former case, the isomerizable carbon-carbon bond becomes completely a single bond, hence determining an energy relaxation on the S_1 potential energy surface. In principle, from this structure the bond can isomerize by twisting in a clockwise or counterclockwise direction (see Figure 2C). In practice, the opsin pocket around the retinal is never symmetric, hence favoring sterically (by a ground state pre-twist) and/or electrostatically (by the charges of the opsin residues) a certain twisting direction. For the same reason, the retinal surrounding can also lead to a partial bond length alternation, determining a S_1 minimum structure: the isomerizable carbon-carbon bond has still some double bond character, hence requiring overcoming a S_1 energy barrier in order to twist (Figure 4A). In any case, the twisting of the isomerizable bond can be accomplished through a S_1/S_0 conical intersection, that functions as a funnel connecting the excited and ground state. Two possibilities arise when passing through the conical intersection: internal conversion, by which the system can revert to its initial state, or further twisting of the isomerizable bond, finally allowing to reach the final state.

The ratio of the number of isomerized versus reverted molecules is the quantum yield of the process, that can reach a high 0.65 value in the process of vision, suggesting the high efficiency by which rhodopsin can initiate the electrical signal cascade. Moreover, the $S_1 \rightarrow S_0$ decay event is expected in *ca.* 110 fs, definitely an ultrafast event that can be completed on the ground state in the sub-ps time scale (Polli et al. 2010).

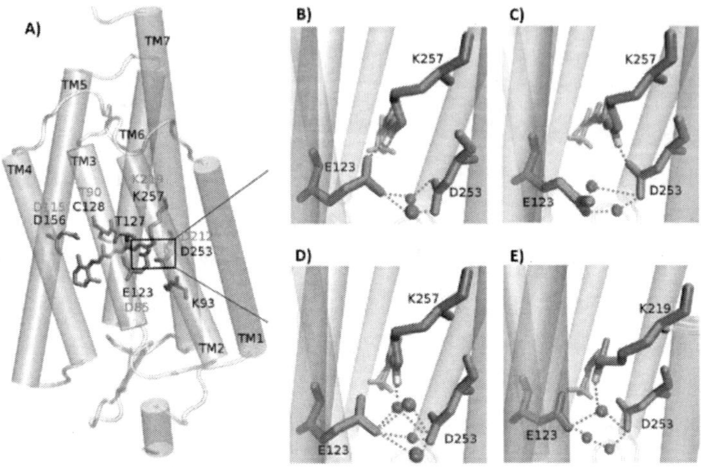

Figure 3. (A) Representation of the opsin pocket formed by the retinal and the most important amino acid residues, written in one-letter code. The residues of ChR2 are written in black, while the residues of bR in grey. The three proposed ChR2 hydrogen bonding patterns are shown as insets, identifying as counterion (B) the negatively charged E123, (C) the negatively charged D253 and (D) a water molecule, bridging E123 and D253 together with other water molecules. (E) The stable three-water pentagonal pattern characteristic of bR is depicted. Adapted from ref. (Guo et al. 2016). Reproduced by permission of The Royal Society of Chemistry.

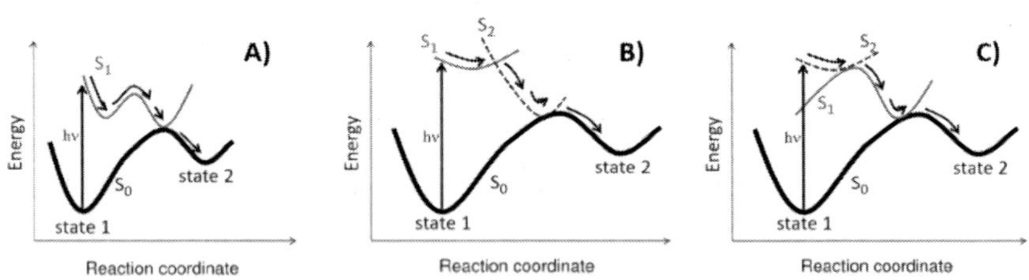

Figure 4. Retinal photoisomerization schemes depicting (A) the two-state model, (B) the three-state model involving absorption to S_1 and (C) the three-state model involving absorption to S_2.

Nonetheless, other types of rhodopsin require a three-state model, comprising S_0, S_1 and S_2 electronic states (Figure 4B, C). In all cases, the adiabatic nuclear potential energy surfaces were calculated until finding intersections (conical intersections or avoided crossings) between two electronic states, where the Born–Oppenheimer approximation is not valid anymore. Actually, Cembran et al. studied a general case, placing in vacuum a negatively charged acetate 3 Ångstrom distant from a protonated Schiff base linked to a polyene chain, thus mimicking a glutamic or aspartic acid side chain as counterion of the retinal, strongly interacting. In this condition, S_2/S_1 crossings were reported to be possibly relevant (Cembran et al. 2004). Indeed, years before Hasson et al. (Hasson, Gai, and Anfinrud 1996) proposed a three-state model of bR: after absorption to an optically bright state (S_1), an intermediate dark state (initially S_2) is populated before crossing with the ground state (Figure 4B), hence determining an adiabatic S_2/S_1 representation, thanks to an avoided crossing that keeps the system constantly on the lower-energy S_1 surface, but allowing to hop from the bright to the dark state. This proposed model was never corroborated experimentally. Alternatively, Luk et al. (Luk et al. 2015; Gozem et al. 2017) proposed a three-state model where the initially populated S_1 (bright state, of charge transfer character) intersects with S_2 (dark state). Nevertheless, in this case the dark state is not expected to be populated, due to a low coupling. Instead, after such intersection, the S_1 potential energy surface continues to be characterized by a charge transfer character, finally intersecting S_0 and achieving the photoisomerized state.

Recently (Guo et al. 2018), different photoisomerization models were proposed for ChR2 depending on the hydrogen bonding pattern: when E123 is the counterion, the two-state model can describe the whole process, S_1 being the charge transfer state, until the S_1/S_0 conical intersection is located (Figure 4A). On the other hand, when a water molecule or D253 is the counterion, a three-state model is required, in both cases necessitating a direct $S_0 \rightarrow S_2$ absorption (Figure 4C). Indeed, the environment destabilizes the charge transfer state, now being S_2. After overcoming a few kcal/mol energy barrier, a S_2/S_1 conical intersection is reached. S_1 is an optically-dark locally-excited state, hence the crossing is inefficient (i.e., a diabatic representation), allowing the charge transfer state to hop on the S_1 state and decrease in energy until a S_1/S_0 conical intersection is found, from where twisting of the isomerizable bond can further proceed. Figure 4 summarizes the possible isomerization models.

3. Biomimetic Retinal Switches

Molecular switches are among the most studied molecular devices due to their function and straightforward applicability. Among them, photoswitches that is, switches activated by light, and in particular the ones based on *E/Z* photoisomerization, have been extensively investigated in the last decades (Russew and Hecht 2010; García-Iriepa, Marazzi, Frutos, et al. 2013). Known examples are azobenzene-like photoswitches (including a large variety of synthesized derivatives and a wide amount of discovered applications), stilbene-like photoswitches and overcrowded alkenes. However, there is still the necessity to design novel families of photoswitches to improve existing properties or propose novel applications.

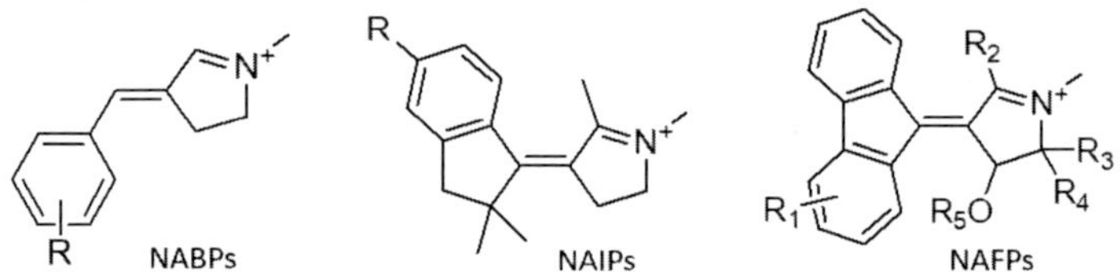

Figure 5. Chemical structure of the three developed generations of retinal-like photoswitches.

In this regard, a worthwhile strategy is the design of photoswitches based on mimicking the elegant and efficient molecular switches found in Nature (Sampedro et al. 2004). As presented in the previous section, the retinal chromophore inside rhodopsin is a clear example of an efficient and ultrafast natural photoswitch due to its outstanding properties: ultrafast photoisomerization together with an exceptionally high quantum yield (Mathies and Lugtenburg 2000). However, these extraordinary properties are worsened when the retinal chromophore is extracted from its opsin pocket and placed in solution, highlighting the great influence of the protein environment (Kandori, Shichida, and Yoshizawa 2001). For this reason, the use of the natural retinal chromophore as a photoswitch did not find appreciable applications out of the rhodopsin field. Nevertheless, the retinal chromophore can be considered as a reference model to build artificial biomimetic switches, whose properties resemble the ones of the retinal *in vivo*.

Following this idea, the first retinal-like photoswitch was reported in 2004 (Sampedro et al. 2004). However, although it served as a promising starting point, its properties as a molecular photoswitch were far from matching the standards of the retinal *in vivo*. For this reason, novel generations of retinal-like photoswitches were proposed with enhanced properties. In this section, we will focus on the three developed generations of retinal-like photoswitches: *i)* *N*-alkylated benzylidene-pyrrolines (NABPs) (Sampedro et al. 2004; Blanco-Lomas, Campos, and Sampedro 2012; Blanco-Lomas et al. 2014), *ii)* *N*-alkylated indanylidene-pyrrolines (NAIPs) (Lumento et al. 2007; Zanirato et al. 2007; A. Sinicropi et al. 2008; Melloni et al. 2010; Briand et al. 2010) and *iii)* *N*-alkylated fluorenylidene-pyrrolines (NAFPs) (Rivado-Casas et al. 2009; Rivado-Casas, Campos, and Sampedro 2010; Rivado-Casas et al. 2011). First, the synthesis of these photoswitches is presented followed by the analysis of their photophysical and photochemical properties.

3.1. Synthetic Pathway

The first generation of biomimetic retinal-like photoswitches was designed taking as reference the retinal chromophore but blocking the double bonds in the pyrroline and benzylidene moieties, except the desired photoisomerizable double bond. Regarding its synthesis, NABPs (**2**) have been obtained by an aldol-like condensation reaction between 1-pyrroline (**1**) and different aldehydes, followed by N-methylation with methyl triflate (Figure 6) (Sampedro et al. 2004). In all cases only the *E* isomer has been obtained. It should be noted that 1-pyrroline presents a high tendency to trimerize. However, it has been reported that the trimer is in equilibrium with

1-pyrroline which in turn reacts with the desired aldehyde (Nomura et al. 1983). The main drawback of this synthetic path is that only three derivatives have been synthetized.

In order to broaden the number of derivatives and in turn, of properties, an alternative synthetic route has been proposed (Blanco-Lomas, Campos, and Sampedro 2012; Blanco-Lomas et al. 2014). Contrary to the previous path, in this case first the photoisomerizable double bond is formed and in the last step, the pyrroline ring is closed (Figure 7). The path starts with the synthesis of the desired phosphonate (**4**), carrying the final R_1 substituent of the pyrroline ring, from the corresponding 2-bromoketone (**3**). Afterwards, an ethyl azide side chain is introduced via reaction of **4** with 2-iodoethylazide, leading the azido-phosphonate **5**. In the next step, the photoisomerizable double bond is formed (**6**) by a Horner-Wadsworth-Emmons reaction (also known as Wittig-Horner reaction) of the azido-phosphonate (**5**) with the desired aldehyde. By intramolecular aza-Wittig reaction with triphenylphosphine, the pyrroline ring is closed (**7**). Finally, NABPs derivatives **8** are obtained by N-methylation with methyl triflate. The main advantage of this synthetic route is that different substituents can be included both in the pyrroline ring and in the benzylidene part, which can be used as anchoring points for various applications.

NABPs constituted the first generation of retinal mimicked photoswitches. In order to improve their properties, still far from the natural ones, a second generation of biomimetic photoswitches emerged known as N-alkylated indanylidene-pyrrolines (NAIPs). The main structural difference between NABPs and NAIPs switches is that the single bond connecting in NABPs the photoisomerizable double bond and the benzylidene moiety, has been integrated in a rigid ring for NAIPs (Figure 5).

Regarding their synthesis, similarly to the alternative path for the synthesis of NABPs (Figure 7), the heterocyclic ring is formed in the last step. The first attempt to synthesize NAIPs derivatives was carried out through a rather long (six steps) and low-yielding path (Figure 8) (Lumento et al. 2007; Zanirato et al. 2007). In detail, starting from 5-methoxy-2,2-dimethyl-1-indanone **9**, via methylation of the corresponding indanone, the cyclopropylcarbinol **10** was obtained by a Grignard addition reaction. Subsequent treatment with HBr and AcOH leads to the formation of the bromopropylidene derivative **11**. In the next two steps the bromide is displaced with sodium azide (**12**) followed by the transformation of the azide group to acetamide (**13**). Finally, the NAIP **15** was obtained after generation of the nitrilium ion by the action of trimethylsilyl polyphosphate (PPSE) used in the Gawley's protocol, followed by N-methylation with methyl triflate. A limitation of this synthetic route is that the α-hydrogen atoms in the C2 position of the starting 5-methoxy-2,2-dimethyl-1-indanone **9** have to be removed in order to obtain the desired exocyclic double bond in **11**. Following this synthetic pathway, a NAIP derivative with a permanent zwitterionic head has been obtained, which allows a reversible photoinduced change on the dipole moment (Melloni et al. 2010).

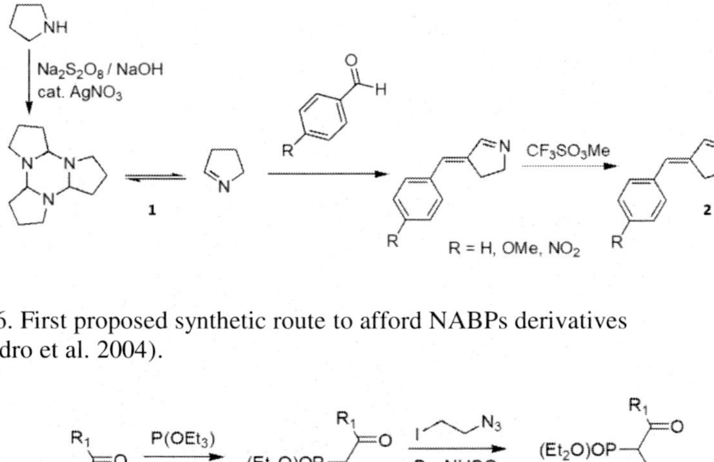

Figure 6. First proposed synthetic route to afford NABPs derivatives (Sampedro et al. 2004).

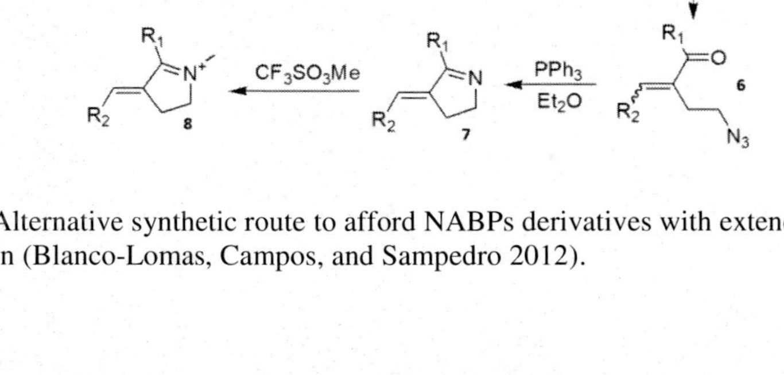

Figure 7. Alternative synthetic route to afford NABPs derivatives with extended substitution (Blanco-Lomas, Campos, and Sampedro 2012).

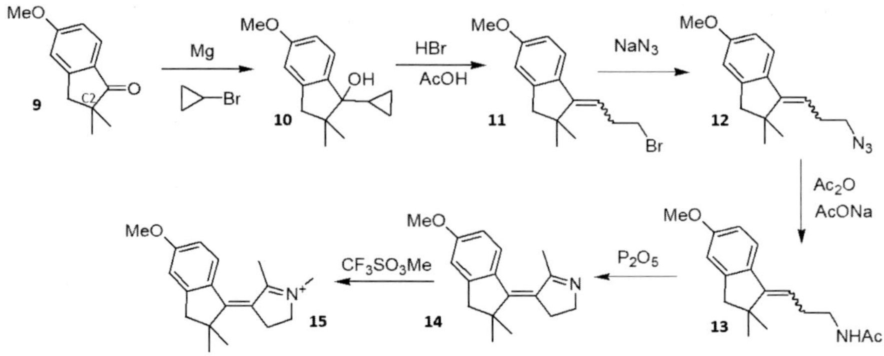

Figure 8. First proposed synthetic route to obtain NAIPs photoswitches (Lumento et al. 2007; Zanirato et al. 2007).

Aimed at improving this synthetic pathway in terms of number of steps and overall yield, an alternative route was proposed using the possibility to generate nitrilium ions, in particular through direct N-alkylation of acetonitrile (Figure 9) (Lumento et al. 2007; Zanirato et al. 2007). To check this hypothesis, the NAIP photoswitch **15** was obtained starting from the cyclopropylcarbinol **10**, synthetized as already described in the previous route (Figure 8), by treatment with acetonitrile and Tf_2O, followed by N-methylation (Figure 9A). In detail, first the cyclopropylcarbinyl cation **I** is formed, evolving via the nitrilium ion **II** and the tertiary-benzylic carbocation **III**. The driving force of this cyclopropyl ring-opening/nitrilium ion ring-closing tandem reaction is the relief of the cyclopropyl ring strain and the extension of the π-system of the backbone. It should be remarked that this synthetic pathway significantly decreases the number of steps up to three and enhances the overall yield.

Moreover, it has been checked if this synthetic route could allow the synthesis of NAIPs derivatives with α-hydrogen atoms at the indanone C2 position, a limitation of the first synthetic path as already mentioned. Indeed, novel NAIPs derivatives **15a** and **15b** were prepared with no methyl groups or only one methyl group in the C2 position, opening the way to chiral NAIPs photoswitches that could act as molecular rotors (Figure 9B) (Zanirato et al. 2007). The versatility of this alternative route has been shown by the synthesis of a NAIP derivative incorporating an unnatural α-amino acid (Melloni et al. 2010).

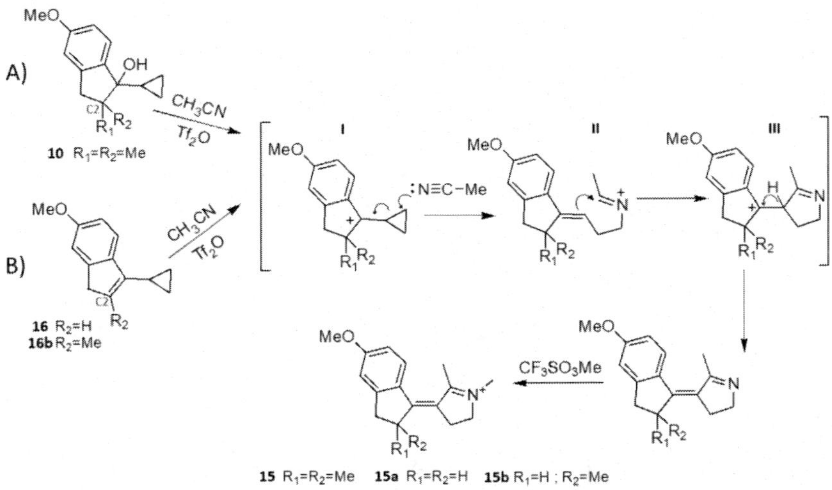

Figure 9. Alternative synthetic route for NAIPs A) without and B) with α-hydrogens in the C2 position of the indanone (Lumento et al. 2007; Zanirato et al. 2007).

The third generation of retinal-like photoswitches are the N-alkylated fluorenylidene-pyrrolines (NAFPs), for which the indanylidene moiety of NAIPs has been replaced by a fluorenylidene unit (Figure 5). This way, the extension of the π-conjugated system is significantly increased, leading to some changes of its photochemical and photophysical properties as will be discussed in the next section.

The synthesis of NAFP derivatives is quite versatile and allows the introduction of different types of substituents in diverse positions of the photoswitch (Figure 10) (Rivado-Casas et al. 2009; Rivado-Casas, Campos, and Sampedro 2010). NAFP derivatives (**20**) are obtained by reaction of fluorene imines (**17**) with alkynyl Fischer carbene complexes (**18**), followed by N-methylation with methyl triflate. The first reaction is the nucleophilic attack of the imine to the alkyne. Different metals such as tungsten or chromium carbene complexes have been tested, obtaining better results for the tungsten compound. It should be noted that the introduction of a R_3 substituent different from a methyl group, induces chirality in the molecule, leading to chiral NAFP photoswitches, a desired property for selected applications.

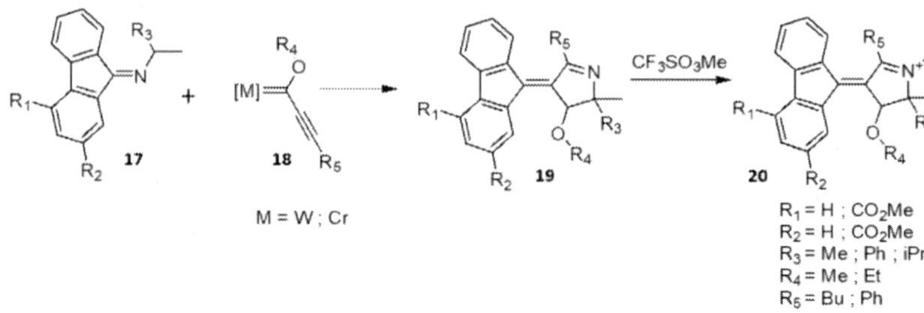

Figure 10. Synthetic route for NAFPs photoswitches (Rivado-Casas et al. 2009; Rivado-Casas, Campos, and Sampedro 2010).

3.2. Photophysical and Photochemical Properties

The three generations of retinal-like photoswitches presented above (NABPs, NAIPs and NAFPs) have some common properties which distinguish them from the other families of *E*/*Z* molecular photoswitches. Concerning the nature of the photoswitch, retinal-based ones present a significant polarity in their backbone due to the presence of an iminium moiety, mimicking the protonated Schiff base of the retinal chromophore (Figure 5).

Regarding their mechanism, the photoisomerization takes place exclusively by the π,π^* excited state (Sampedro et al. 2004; Lumento et al. 2007; A. Sinicropi et al. 2008), contrarily to azobenzene-like photoswitches for which the isomerization can occur involving both π,π^* or n,π^* states (Siampiringue et al. 1987). Hence, this simplified mechanism allows more control on the photoswitch properties and facilitates the computational study. Concerning the photoisomerization mechanism (Sampedro et al. 2004; Lumento et al. 2007; A. Sinicropi et al. 2008), the system follows mainly the scheme presented in Figure 11: after excitation to the lowest-lying singlet excited state (π,π^*), the system activates mainly the stretching mode that is, bond length alternation of single and double bonds, reaching a shallow minimum with a limited torsional deformation. From this minimum, the system evolves along the torsion mode by twisting the photoisomerizable carbon-carbon bond. Hence, an energy barrier needs to be overcome, usually corresponding to only a few kcal/mol due to the flat potential energy surface along this coordinate. Descending in energy, a conical intersection between the first excited state and the ground state is located, allowing the system to decay non-radiatively to the ground state and either complete the photoisomerization or revert to the initial isomer, i.e., internal conversion (Figure 11).

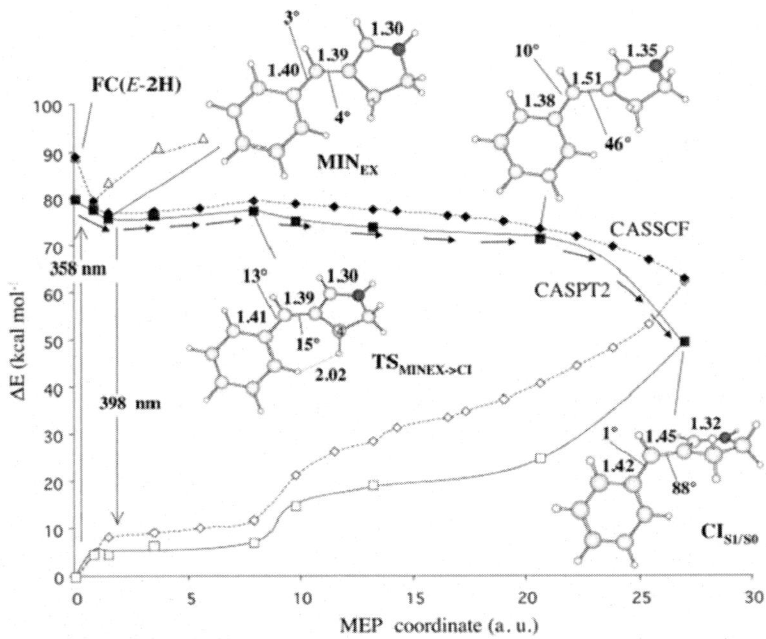

Figure 11. Minimum energy paths (MEPs) computed at the CASSCF and CASPT2//CASSCF levels of theory for the first reported NABPs derivative, from $S_0 \rightarrow S_1$ vertical excitation to S_1/S_0 conical intersection. Reprinted with permission from (Sampedro et al. 2004). Copyright year 2004 American Chemical Society.

Another common property of these three generations of retinal-like photoswitches is the charge transfer character of the involved excited state. In the ground state, the positive charge is located in the iminium moiety. Upon vertical excitation, the charge is transferred through the central and photoisomerizable double bond to the aromatic unit, where the charge is delocalized (Blanco-Lomas et al. 2012; Lumento et al. 2007). This finding is similar to the excited state charge transfer experienced by the retinal chromophore inside rhodopsin (as also stated in section 2), with the amino acids' side chains of the binding pocket aiding in stabilizing the positive charge away from the iminium group.

As suggested by the computed path, characterized by an excited state shallow minimum, the photoisomerization of retinal-like photoswitches is an ultrafast process. For instance, the photoisomerization of NAIPs switches takes place on a 0.3 ps time scale, followed by <10 ps cooling and solvation (A. Sinicropi et al. 2008). Moreover, the full $Z \rightarrow E \rightarrow Z$ cycle is completed within 20 ps (A. Sinicropi et al. 2008). These time scales are similar to the ones obtained for retinal *in vivo* (Mathies and Lugtenburg 2000). It is expected that the excited state lifetime and isomerization timescale of NABPs and NAFPs are similar to the ones recorded for NAIPs, as their mechanism also point towards an ultrafast process but, unfortunately, these data have not been measured for the corresponding derivatives. It should be remarked that for NAIPs derivatives, the excited state decay is slowed down when the methyl group close to the iminium moiety is eliminated. In this case, the system relaxes in 500 ps, being almost a factor of two longer (Dunkelberger et al. 2012). This finding has been reasoned on the basis of smaller steric hindrance between the two moieties connected by the photoisomerizable double bond, due to the suppression of the methyl group, leading to a more planar geometry (Hu et al. 2015). Thus, the system reaches the conical intersection at larger twisting angles, hence slowing down the excited state decay kinetics.

In order to find possible applications, essential properties are photostability coupled to fatigue resistance. This means that the photoswitch can be interconverted between *on* and *off* states along multiple cycles (i.e., fatigue resistance) without undergoing photo-degradation (i.e, photostability). In general, the three generations of retinal-like photoswitches fulfill these criteria (Blanco-Lomas, Campos, and Sampedro 2012; Sampedro et al. 2004; Rivado-Casas et al. 2011). In addition, another common property of these three families is the influence of both excitation wavelength and structural modifications to the composition of the photostationary state (PSS). Especially, for NABPs and NAIPs the *E*/*Z* ratio achieved after irradiation can be tuned by changing the excitation wavelength (Lumento et al. 2007; Blanco-Lomas et al. 2012). For instance, for a NAIPs derivative the PSS is enriched of the *Z* isomer when exciting at 360 nm, but it is inverted when irradiating at 400 nm, finding the *E* isomer as the major photoproduct (Lumento et al. 2007). This fact can be explained by the different absorption of both the *E* and *Z* isomers at the different wavelengths: when the *E* isomer mainly absorbs at the selected irradiation wavelength, the PSS will be enriched of the *Z* isomer, and vice versa. Also, it has been found for the presented three retinal-like generations that the PSS depends on the photoswitch derivative, that is on the different chemical substituents. Hence, retinal-like photoswitches allow tuning the PSS composition by a selected combination of both, the excitation wavelength and the substituents pattern.

Although the three generations of retinal-like photoswitches present several common properties, they differ in key features determining their distinct performance, that increases from generation to generation. For the first generation (NABPs photoswitches) the main drawback was the quite low photoisomerization quantum yield: $6 \cdot 10^{-3}$ (Sampedro et al. 2004). This means that a reactive path different from photoisomerization is taking place, hence quenching the photoisomerization. For instance, a radiative pathway could dominate such as fluorescence decay. However, the fluorescence quantum yield was measured to be quite low, close to the instrument limit. This means that a non-radiative path should be in competition with the photoisomerization. By analyzing the structure, it was concluded that the rotation of the single bond connecting pyrrolone and benzylidene moieties is most probably the unreactive channel quenching the photoisomerization of NABPs. In order to enhance this property, the NAIPs derivatives emerged. Following the reasoning given for the low photoisomerization quantum yield of NABPs derivatives, NAIPs were designed with the aim to decrease the number of torsional degrees of freedom, in order to avoid competitive pathways. In particular, the single bond free to rotate in NABPs was included in the aromatic moiety of NAIPs. This way, only the rotation around the photoisomerizable double bond is possible. Indeed, the photoisomerization quantum yield was significantly increased until 0.20 and 0.34 for the *Z*→*E* and *E*→*Z* paths, respectively (Lumento et al. 2007). Although the photoisomerization quantum yield has been significantly enhanced, still one feature of NAIPs remains far from desired, that is the photoswitch activation with visible light. In particular, NAIPs can be activated with near-UV light, showing absorption maxima around 390 nm. This goal was reached with the third generation of retinal-like photoswitches (NAFPs) by increasing the π-conjugation in order to red-shift the absorption spectrum. In particular, the indanylidene unit was replaced by a fluorenylidene moiety, achieving a significant red-shift with respect to the NAIPs derivatives (Rivado-Casas et al. 2009, 2011). As a result, NAFPs can be activated by visible light irradiation. Moreover, the photoisomerization quantum yield of these derivatives is 0.5, hence even larger than the one found for NAIPs and getting closer to the quantum yield of retinal within rhodopsin (0.65).

4. Photo-Control of Peptides Secondary Structure

Molecular photoswitches are among the molecular devices with a wider range of applications, from material science to biology. In particular, the photoactive component is integrated into a macromolecular system, inducing a significant change on the properties of the system upon light irradiation. In this regard, *E/Z* photoswitches have been used to photo-control the secondary structure (and hence the biological activity) of a given peptide (Yu and Hecht 2016; Szymański et al. 2013). More in detail, one state of the photoswitch (the *E* or *Z* isomer) is linked to the peptide under study. After irradiation, the photoswitch isomerizes and the concomitant change of its structure leads to the change of the peptide secondary structure. Azobenzene-like photoswitches are by far the most used *E/Z* photoswitches in this field (Renner and Moroder 2006), even though also stilbene (Erdélyi,

Karlén, and Gogoll 2005; Erdélyi et al. 2008) and hemithioindigo-like photoswitches (Cordes et al. 2009; Regner et al. 2012) have found some applications. Regarding the peptide, photoswitches have been linked to both α-helical and β-hairpin structures (Szymański et al. 2013). In addition, different approaches have been used to link the photoswitch to the peptide: *i)* the side chain strategy, with the photoactive moiety covalently linked to the side chain of a residue (Figure 12A) and *ii)* the loop approach, with the photoswitch bridging the peptide (Figure 12B).

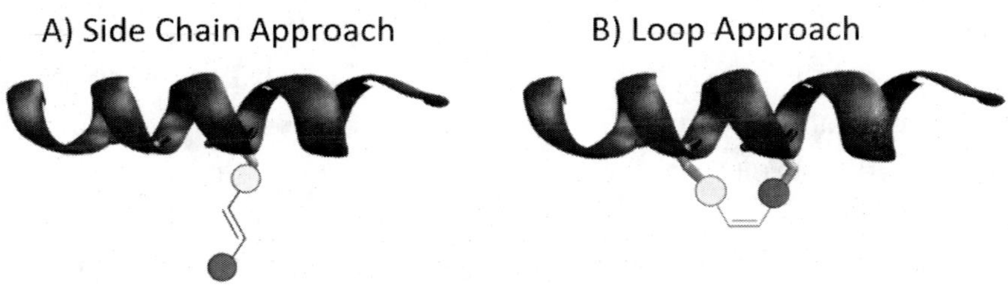

Figure 12. Graphical representation of the photoswitch-peptide linking approaches: A) side chain approach and B) loop approach.

In order to broaden the possibilities of photo-controlling the secondary structure of a given peptide, alternatives to azobenzene-like photoswitches were proposed. Retinal-like photoswitches present properties quite different from azobenzene-like derivatives such as a larger polarity and an increased end-to-end distance. Moreover, as discussed in the previous section, the absorption wavelength, the photoisomerization quantum yield or the photostationary state, can be modulated depending on the derivative. It is remarkable the fact that for some retinal-like photoswitches, visible light can be used to activate the switch, a crucial issue for biological and biomedical applications. Especially, a shift towards the far-red/near-infrared region of the spectrum, thus reaching the so-called therapeutic window (650-1350 nm) would be highly beneficial.

In the next subsections, an overview of the reported photo-controlled secondary structure of certain peptides through retinal-like photoswitches are described. As a first attempt, computational studies were performed for the design and study of the photoswitching mechanism of retinal-like switches linked to peptide minimum models. Then, the experimental evidence of photo-controlling the conformation of an α-helical peptide with a retinal-like switch is presented, supported by computational data.

4.1. Peptide Minimum Models

The first example of a photo-responsive peptide through a retinal-like photoswitch was published in 2005 (Andruniów et al. 2005), selecting an octapeptide derived from the active site of the thioredoxin reductase (Figure 13A) and using the loop approach for the switch linkage. The computational study covers the design, the analysis of the switch-coupled-peptide relaxation and the intrinsic factors driving the transmission of the switch strain to the peptide. By quantum mechanics/molecular mechanics (QM/MM) calculations it has been shown that the switch mechanism is quite similar as in bare solution: after excitation, the stretching coordinate is activated and a minimum in S_1 is reached. By overcoming an energy barrier (in some cases only a few kcal/mol) along the torsion coordinate, a conical intersection with the ground state is reached, decaying to S_0 and completing the isomerization or reverting to the initial conformer. By checking the peptide conformation before excitation and at the conical intersection structure, it has been demonstrated that they are quite similar, and that the change of the peptide structure (hydrogen bonding pattern, side chain orientation, helical character) takes place only after decay to the ground state through the conical intersection. Hence, the relaxation of the switch and the peptide should take place in quite different time scales. To clarify this hypothesis, Car–Parrinello molecular dynamics (CPMD) (Car and Parrinello 1985) trajectories were performed. By analyzing the data, it was shown that the relaxation of the photoswitch is a fast process while the change of the peptide secondary structure takes place in a longer time scale (> 3ps). Moreover, this study demonstrated that the transmission of the switch strain to the peptide follows the sequence switch → link → peptide and that it is asymmetric as the two link regions relax in an asymmetric way. Hence, thanks to this example it was shown the feasibility of using a retinal-like switch to photo-control the secondary structure of a peptide.

A second model example aiming to photo-control the peptide secondary structure using a retinal-like switch appeared in 2009 (Adalgisa Sinicropi et al. 2009). In this case, a NAIP derivative and the Arg-Gly-Asp (RGD) tripeptide were selected as the photoactive and the biological motifs, respectively (Figure 13B). The NAIP derivative was linked to the peptide following the loop approach (Figure 12B). Regarding the RGD peptide, this sequence was explicitly chosen since it plays a role in cell adhesion (Bellis 2011). For this reason, the linkage of the switch to this motif and its photoisomerization could lead to a change of the peptide backbone conformation, allowing the photo-modulation of cell adhesion. The RGD motif can be converted in a cyclic system through a bicyclic lactam bridging the RGD terminations. In this computational study, the bicyclic lactam has been replaced by the photoactive NAIP derivative (Figure 13B). By performing QM/MM calculations, the photoswitching mechanism of the switch linked to the peptide was elucidated. As observed in the previous model system, the mechanism found was similar to the NAIPs bare in solution. Moreover, molecular dynamics (MD) simulations were performed to analyze the effect of the switch photoisomerization on the peptide conformation: the peptide backbone shows changes in the values of the dihedral angles when linked to the *Z* or the *E* isomer. Hence, this study further evidences that retinal-like photoswitches can be used to photo-control the secondary structure of a given peptide.

The study of these peptide minimal models and the demonstration of the photoisomerization efficiency of the retinal-like switches linked to the peptide, opened the way to real applications. In the next section it is presented the first experimental evidence, supported by computational data, of the peptide secondary structure change by a retinal-like switch photoisomerization.

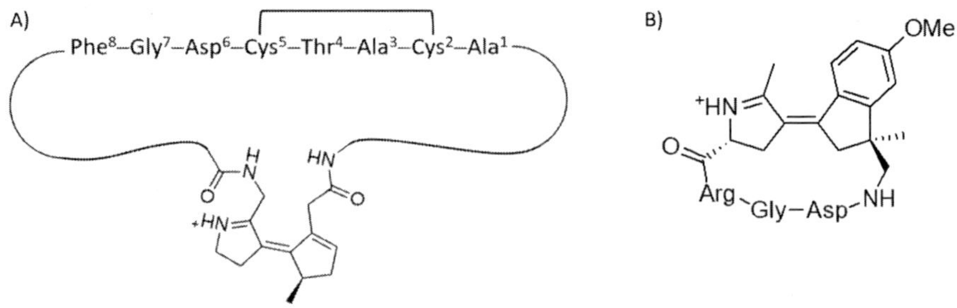

Figure 13. Representation of the two peptide minimal models with a retinal-like photoswitch linked to A) an octapeptide (Andruniów et al. 2005) and B) a tripeptide (Adalgisa Sinicropi et al. 2009).

4.2. An α-helix as a Biochemical Case

After the design and computational studies of peptide minimal models cross-linked to a retinal-like photoswitch, and the demonstration of the feasibility to photo-control its secondary structure, in 2012 appeared the first experimental evidence (Blanco-Lomas et al. 2012). In this particular case, an α-helical peptide of 27 amino acids, known as SS-11, was selected for this study. Again, the loop approach was applied to link the switch, which in this case is a NABP derivative (Figure 14A). More in detail, the photoswitch is covalently linked to the peptide through the sulfur (S) atom of two cysteine (C) residues spaced by 11 amino acids, hence its name SS-11. The distance between the two sulfur atoms matches with the end-to-end distance of the *E* isomer, while the shorter distance characteristic of the *Z* isomer leads to a change of the secondary structure.

After the synthesis of the cross-linked peptide, the photochemical properties were measured, and the secondary structure change of the peptide analyzed by circular dichroism (CD) spectroscopy. Concerning the photophysical and photochemical properties, it was found an efficient photoisomerization of the switch (quantum yield ~ 0.24). Moreover, visible light can be used to photo-activate the *E* isomer. This system is photostable and presents a significant fatigue resistance as multiple rounds of photoisomerization alternating visible and ultra-violet light were performed without observing sample decomposition. Regarding the secondary structure of the peptide, linking the *E* isomer leads to a stabilization of the α-helical content, compared to SS-11 alone. Moreover, upon photoisomerization to the *Z* isomer, a clear decrease in helicity takes place. This conformational change is fully reversible.

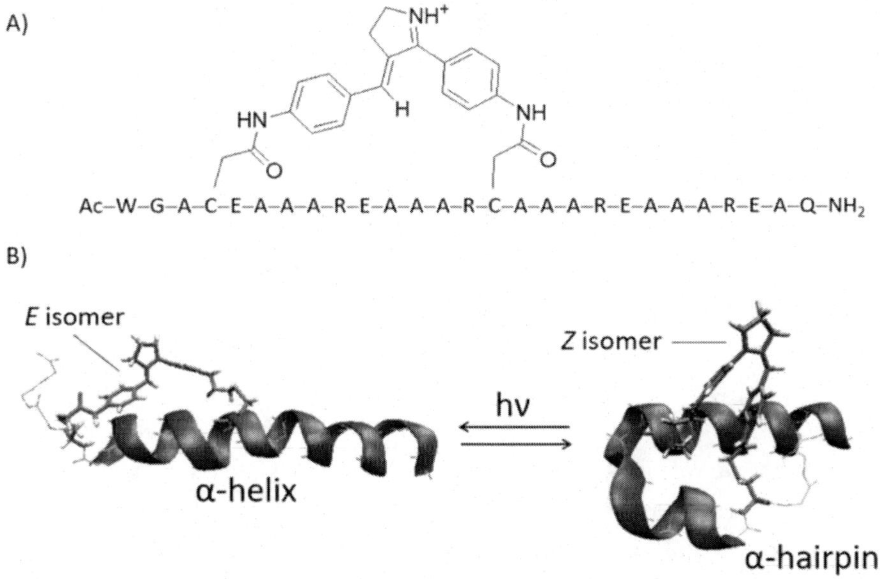

Figure 14. A) Cross-linked SS-11 peptide with a retinal-like switch in its *E* form. B) Peptide conformational change from α-helix to α-hairpin upon switch photoisomerization (García-Iriepa et al. 2016).

To get insight into these findings, a computational study (MD simulations and QM/MM calculations) together with transient absorption spectroscopy (TAS) were performed (García-Iriepa et al. 2016). Two main questions were raised: *i)* the effect of linking the switch to the peptide on its photochemical properties, compared to the switch alone in solution and *ii)* the conformational change of the peptide upon photoisomerization, including the determination of the residues responsible of the helicity decrease.

Regarding the photochemistry of the switch alone in solution and linked to the peptide, the isomerization mechanism is similar for these two systems:

after excitation the stretching coordinate is activated until an excited state minimum is reached. From there, an energy barrier is overcome along the torsion coordinate until the conical intersection with the ground state is found. However, a main difference was found, concerning the excited state decay measured by TAS, being monoexponential for the switch alone in solution and biexponential when linked to the peptide. By analyzing in detail the MD simulations of the *E* isomer linked to the peptide, two main ground state conformations were found, hence explaining the biexponential decay as compared to the switch alone in solution, where a single conformation is responsible of the monoexponential decay. Indeed, the photoisomerization path of the two characteristic conformations of the peptide linked to the switch were computed, finding different energy barriers in the excited state. Moreover, the energy barriers computed are larger than the one of the switch alone in solution. This finding supports the increase found by TAS of up to one order of magnitude in the excited state time decay of the switch, once linked to the peptide.

Regarding the second proposed question, i.e., secondary structure change, the answer is based on the analysis of the MD simulations of different systems: the SS-11 peptide alone and the cross-linked SS-11 peptide with both *E* and *Z* isomers (*E*-SS-11 and *Z*-SS-11, respectively). By comparing the helical content of SS-11 and *E*-SS-11, a higher helicity content was found when the switch is linked, as experimentally observed. In addition, the secondary structure change of the peptide upon photoisomerization was investigated by comparing *E*-SS-11 and *Z*-SS-11. Interestingly, it was found that *E*-SS-11 is an α-helix while *Z*-SS-11 is an α-hairpin (Figure 14B). This conformational change agrees with the decrease in helicity measured by CD spectroscopy after switch irradiation. In order to validate the MD results, the CD spectra of *E*-SS-11 and *Z*-SS-11 were simulated using these dynamics data (Gattuso et al. 2017). Similar CD spectra compared to the experiment were obtained, validating the α-helix to α-hairpin conformational change.

Hence, the use of retinal-like switches to photo-control the secondary structure of peptides has been demonstrated with both experimental and computational studies. Further studies could be proposed to use biological active peptides, hence possibly reaching the photo-control of biological functions (de Castro and Sato 2015; J. Kastin and Pan 2010). It is remarkable the fact that a wide range of retinal-like photoswitches are available (NABPs, NAIPs and NAFPs), allowing the modulation of the desired photophysical and photochemical properties. Moreover, the mechanism of this family of photoswitches is simpler with respect to azobenzene, as only the S_1 (π,π^*) state is involved in the photoisomerization process. This fact facilitates the study and rational analysis of the photoisomerization mechanism of cross-linked peptides.

5. From a Retinal-Based Switch to a Motor

As shown in Figure 2C, a chromophore can be designed as a light-induced rotary motor: the incoming photon energy can be conveniently converted into the rotation – by photoisomerization – of a part of a molecule (the rotor) with respect to the inert rest of the molecule (the stator). This implies that, in order to be effective, multiple photoisomerization processes should take place by twisting always in the same direction, clockwise or counterclockwise. Hence, a retinal-based switch can be converted into a motor if a significant isomerization directionality can be accomplished, leading to complete (360 degrees) and consecutive rotations around the isomerizable carbon-carbon bond. This goal can be performed in different ways, as we describe in the following sections.

5.1. Stereospecific Photoisomerization

The first light-induced rotary molecular motors were synthesized in the late 1990s, based on sterically overcrowded alkenes (Koumura et al. 1999). Different applications were found in order to profit from the photogenerated mechanical motion (Chiang et al. 2012; London et al. 2013; Van Dijken et al. 2016; Chen, Wezenberg, and Feringa 2016; Saywell et al. 2016; Zhao et al. 2017; García-López et al. 2017). Nevertheless, this first generation required an ultrafast stereospecific photochemical step followed by a thermal step in order to rotate 180 degrees. Hence, in total two photochemical and two thermal steps are required to complete a 360 degrees turn, being definitely the thermal steps the limiting ones in terms of kinetics (Figure 15). Moreover, the photoisomerization quantum yields are rather low, ≤ 0.3. This is mainly due to an undesired coupling of the torsional motion with the pyramidalization of a carbon atom (Kazaryan et al. 2011; Filatov and Olivucci 2014). Hence, a strong effort was directed toward the synthesis and characterization of molecular motors with accelerated thermal steps, improving the properties of sterically overcrowded alkenes (Pérez-Hernández and González 2010; Fang, Oruganti, and Durbeej 2014; Oruganti, Fang, and Durbeej 2015; Oruganti, Wang, and Durbeej 2016; Bauer et al. 2014; Vachon et al. 2014; Faulkner et al. 2016).

Alternatives were also proposed to eliminate the thermal steps, hence obtaining a full rotation of 360 degrees by two consecutive C=C photoisomerizations. Especially, in 2013 a chiral hydrogen bond environment was proposed to generate efficient molecular motors based on retinal. Especially, two chiral carbon atoms are present, each of them connected to a OH group, thus permitting the formation of hydrogen bonds with the protonated Schiff base (García-Iriepa, Marazzi, Zapata, et al. 2013) (Figure 16A). Two years later, it was shown that placing a stereogenic center in an allylic position with respect to the isomerizing C=C bond of a retinal-based switch would also result in a significant photoisomerization directionality, also eliminating the two thermal steps and requiring two consecutive photoisomerization steps (Marchand et al. 2015) (Figure 16B). Indeed, the stereogenic center is able to create an asymmetry of the excited state force field, that results in a preferred (*Z*→*E* or *E*→ *Z*) conversion. This ensures a given directionality in the photoisomerization process (>70%). Such excited state directionality is guaranteed also by the chiral hydrogen bond environment approach. Nevertheless, while the intramolecular hydrogen bond formed by applying the chiral hydrogen bond environment would ensure also a considerable pre-twist of the isomerizable bond on the ground state, this is not the case for the single stereogenic center approach, by which the motor is almost planar before photon absorption.

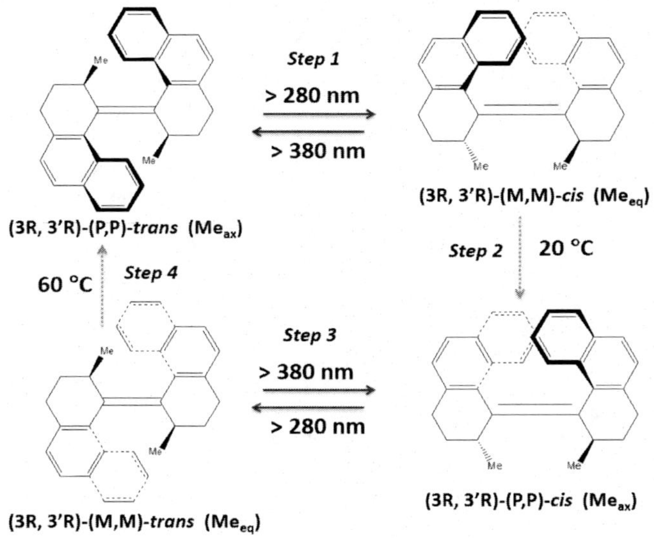

Figure 15. Scheme depicting the four necessary steps of a sterically overcrowded alkene to complete one turn (360 degrees) by twisting unidirectionally the carbon-carbon double bond connecting the two moieties. Steps 1 and 3 are the fast photochemical steps, being possible for each of them to revert the sense of rotation by applying light-sources of different energy (280 and 380 nm). Steps 2 and 4 are the slow thermal steps, being necessary to heat at 20 and 60 °C, respectively.

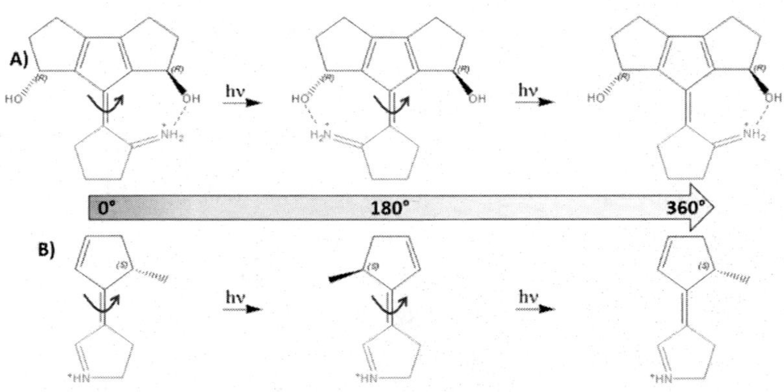

Figure 16. Proposed molecular motors based only on two photochemical steps, derived from a protonated Schiff base: A) a chiral hydrogen bond environment is generated by two stereocenters that promote the formation of O···H bonds and B) chirality induced by a single stereocenter. In both cases, absorption of two photons of equivalent energy produces a complete 360 degrees turn. As for the sterically overcrowded alkenes, only one double bond – connecting stator and rotor moieties – is allowed to isomerize.

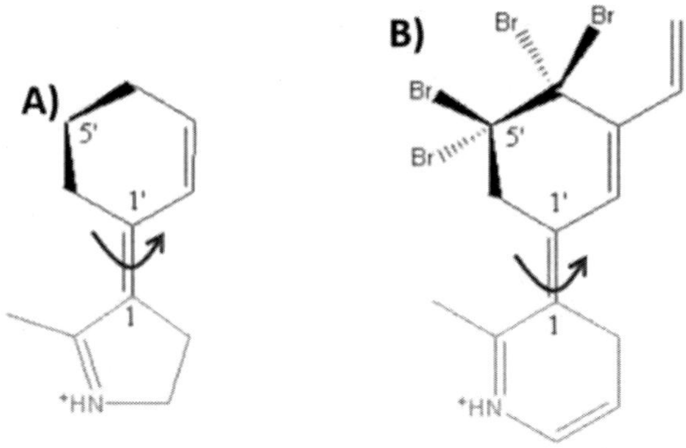

Figure 17. Alternative molecular motors based on protonated Schiff bases, not necessitating a stereocenter: A) minimal model and B) modified compound absorbing in the visible part of the spectrum. In both cases, the asymmetry of the puckered ring (position 5') is driving the photoisomerization direction.

5.2. Molecular Motors without a Stereocenter

Recently, retinal-based molecular motors were engineered in order to demonstrate that point chirality is not an essential feature of rotary molecular motors based on photoisomerization. Indeed, it is possible to benefit from the intrinsic asymmetry of a puckered organic cycle to control the unidirectionality of the rotary motion. Especially, a ciclohexenylidene moiety was introduced in a protonated Schiff base framework, demonstrating that the orientation of the cycle can induce a light-activated unidirectional rotation (Wang, Oruganti, and Durbeej 2017) (Figure 17A). Moreover, these novel molecular motors were additionally modified in order to meet another property recently challenged: activation by visible light. Until now, highly extended π-conjugated systems and push-pull overcrowded alkenes' substituents allowed to reach the violet and blue part of the visible spectrum (Van Leeuwen et al. 2017; Van Delden et al. 2003). Consistent red-shifts were obtained only by integrating metal centers, by porphyrin systems or organometallic complexes, that served as photosensitizers' agents (Cnossen et al. 2012; Wezenberg, Chen, and Feringa 2015). In all cases, these motors required also thermal steps. On the other hand, it was demonstrated by non-adiabatic molecular dynamics that a modified ciclohexenylidene moiety – through the inclusion of Br substituents – coupled to extended conjugation of the protonated Schiff base fragment can also allow to reach the visible range (Wang and Durbeej 2018) (Figure 17B).

CONCLUSION

In this chapter, we have shown how the isomerization of a carbon-carbon double bond is at the basis of highly relevant biological functions, as vision in mammalians and ion-channels for phototaxis. This is possible through a photo-receptor called retinal that, coupled to an opsin pocket to form a transmembrane rhodopsin protein, transduces incoming light into an electrical signal. The first step of this cascade of reactions is the photoisomerization of retinal. After having discussed the characteristics of the different types of rhodopsin, a detailed explanation was given concerning the various photoisomerization mechanisms, depending on the biological environment. The knowledge obtained by studying rhodopsins was a source of inspiration for several research groups, in order to mimic the retinal *in vivo* exceptional properties and propose bio-inspired photoswitches. All possible synthetic pathways were carefully addressed, as well as the photophysical and photochemical properties, by both experiment and theory when available. Furthermore, the capability of such retinal-based photoswitches to be applied in the biochemical field was demonstrated by photo-controlling the secondary structure of short peptides until a 27 amino acids' α-helix. In this last case, the two states of the photoswitch correspond to two different secondary structures, making possible a novel interconversion between an α-helix and an α-hairpin.

In the last section, the attention is focused on photoinduced molecular motors, by considering the photoisomerizable double bond as an engine converting light into unidirectional rotary motion. In this way, a rotor and a stator coupled to the engine can constitute the molecular motor. Especially all generations of motors, from the sterically overcrowded alkenes until the newest organic minimal models not necessitating stereocenters, were shown. Especially, we highlight how retinal-based molecular motors can definitely improve the rotation speed, since only photochemical and no thermal steps are required, in contrast to the sterically overcrowded alkenes. Still, synthetic issues need to be properly addressed in order to guarantee their practical use. Also, visible incoming light can be now applied through different techniques, including photosensitization and organic substitutions, finally envisaging applications in the biological and biomedical field, as it is the case of photoswitches.

References

Agathangelou, D., Y. Orozco-Gonzalez, M. Del Carmen Marín, P. P. Roy, J. Brazard, H. Kandori, K. H. Jung, et al. 2018. "Effect of Point Mutations on the Ultrafast Photo-Isomerization of Anabaena Sensory Rhodopsin." *Faraday Discussions* 207: 55–75. doi:10.1039/c7fd00200a.

Altoe, P., A. Cembran, M. Olivucci, and M. Garavelli. 2010. "Aborted Double Bicycle-Pedal Isomerization with Hydrogen Bond Breaking Is the Primary Event of Bacteriorhodopsin Proton Pumping." *Proceedings of the National Academy of Sciences* 107 (47): 20172–77. doi:10.1073/pnas.1007000107.

Andruniow, T., N. Ferre, and M. Olivucci. 2004. "Structure, Initial Excited-State Relaxation, and Energy Storage of Rhodopsin Resolved at the Multiconfigurational Perturbation Theory Level." *Proceedings of the National Academy of Sciences* 101 (52): 17908–13. doi:10.1073/pnas.0407997101.

Andruniów, Tadeusz, Simona Fantacci, Filippo De Angelis, Nicolas Ferré, and Massimo Olivucci. 2005. "Mechanism of the Initial Conformational Transition of a Photomodulable Peptide." *Angewandte Chemie - International Edition* 44 (37): 6077–81. doi:10.1002/anie.200501145.

Balashov, S P, and J K Lanyi. 2007. "Xanthorhodopsin: Proton Pump with a Carotenoid Antenna." *Cellular and Molecular Life Sciences : CMLS* 64 (18): 2323–28. doi:10.1007/s00018-007-7167-y.

Balzani, Vincenzo, Alberto Credi, and Margherita Venturi. 2009. "Light Powered Molecular Machines." *Chemical Society Reviews*. doi:10.1039/b806328c.

Bauer, Jurica, Lili Hou, Jos C. M. Kistemaker, and Ben L. Feringa. 2014. "Tuning the Rotation Rate of Light-Driven Molecular Motors."*Journal of Organic Chemistry* 79 (10): 4446–55. doi:10.1021/ jo500411z.

Bellis, Susan L. 2011. "Advantages of RGD Peptides for Directing Cell Association with Biomaterials." *Biomaterials* 32 (18): 4205–10. doi:10.1016/j.biomaterials.2011.02.029.

Berthold, P., S. P. Tsunoda, O. P. Ernst, W. Mages, D. Gradmann, and P. Hegemann. 2008. "Channelrhodopsin-1 Initiates Phototaxis and Photophobic Responses in Chlamydomonas by Immediate Light-Induced Depolarization." *The plant cell online* 20 (6): 1665–77. doi:10.1105/tpc.108.057919.

Blanco-Lomas, Marina, Pedro J. Campos, and Diego Sampedro. 2012. "Synthesis and Photoisomerization of Rhodopsin-Based Molecular Switches." *European Journal of Organic Chemistry*, no. 32: 6328–34. doi:10.1002/ejoc.201200870.

Blanco-Lomas, Marina, David Martínez-López, Pedro J. Campos, and Diego Sampedro. 2014. "Tuning of the Properties of Rhodopsin-Based Molecular Switches." *Tetrahedron Letters* 55 (22): 3361–64. doi:10.1016/j.tetlet.2014.04.054.

Blanco-Lomas, Marina, Subhas Samanta, Pedro J. Campos, G. Andrew Woolley, and Diego Sampedro. 2012. "Reversible Photocontrol of Peptide Conformation with a Rhodopsin-like Photoswitch." *Journal of the American Chemical Society* 134 (16): 6960–63. doi:10.1021 /ja301868p.

Briand, Julien, Olivier Bräm, Julien Réhault, Jérémie Léonard, Andrea Cannizzo, Majed Chergui, Vinizio Zanirato, Massimo Olivucci, Jan Helbing, and Stefan Haacke. 2010. "Coherent Ultrafast Torsional Motion and Isomerization of a Biomimetic Dipolar Photoswitch."

Physical Chemistry Chemical Physics 12 (13): 3178–87. doi:10.1039/b918603d.

Browne, Wesley R., and Ben L. Feringa. 2006. "Making Molecular Machines Work." *Nature Nanotechnology* 1 (1): 25–35. doi:10.1038/nnano.2006.45.

Car, R., and M. Parrinello. 1985. "Unified Approach for Molecular Dynamics and Density-Functional Theory." *Physical Review Letters* 55 (22): 2471–74. doi:10.1103/PhysRevLett.55.2471.

Castro, Ruann Janser Soares de, and Hélia Harumi Sato. 2015. "Biologically Active Peptides: Processes for Their Generation, Purification and Identification and Applications as Natural Additives in the Food and Pharmaceutical Industries." *Food Research International*. doi:10.1016/j.foodres.2015.05.013.

Cembran, Alessandro, Fernando Bernardi, Massimo Olivucci, and Marco Garavelli. 2004. "Counterion Controlled Photoisomerization of Retinal Chromophore Models: A Computational Investigation." *Journal of the American Chemical Society* 126 (49): 16018–37. doi:10.1021/ja048782+.

Chen, Jiawen, Sander J. Wezenberg, and Ben L. Feringa. 2016. "Intramolecular Transport of Small-Molecule Cargo in a Nanoscale Device Operated by Light." *Chemical Communications* 52 (41): 6765–68. doi:10.1039/c6cc02382g.

Chiang, Pinn Tsong, Johannes Mielke, Jazmin Godoy, Jason M. Guerrero, Lawrence B. Alemany, Carlos J. Villagómez, Alex Saywell, Leonhard Grill, and James M. Tour. 2012. "Toward a Light-Driven Motorized Nanocar: Synthesis and Initial Imaging of Single Molecules." *ACS Nano* 6 (1): 592–97. doi:10.1021/nn203969b.

Cnossen, Arjen, Lili Hou, Michael M. Pollard, Philana V. Wesenhagen, Wesley R. Browne, and Ben L. Feringa. 2012. "Driving Unidirectional Molecular Rotary Motors with Visible Light by Intra- and Intermolecular Energy Transfer from Palladium Porphyrin." *Journal of the American Chemical Society* 134 (42): 17613–19. doi:10.1021/ja306986g.

Cordes, Thorben, Cord Elsner, Teja T. Herzog, Christian Hoppmann, Torsten Schadendorf, Wolfram Summerer, Karola Rück-Braun, and Wolfgang Zinth. 2009. "Ultrafast Hemithioindigo-Based Peptide-Switches." *Chemical Physics* 358 (1–2): 103–10. doi:10.1016/j.chemphys.2008.12.027.

Delden, Richard A. Van, Nagatoshi Koumura, Annemarie Schoevaars, Auke Meetsma, and Ben L. Feringa. 2003. "A Donor-Acceptor Substituted Molecular Motor: Unidirectional Rotation Driven by Visible Light." *Organic and Biomolecular Chemistry* 1 (1): 33–35. doi:10.1039/b209378b.

Dijken, Derk Jan Van, Jiawen Chen, Marc C. A. Stuart, Lili Hou, and Ben L. Feringa. 2016. "Amphiphilic Molecular Motors for Responsive Aggregation in Water." *Journal of the American Chemical Society* 138 (2): 660–69. doi:10.1021/jacs.5b11318.

Dokukina, Irina, Artur Nenov, Christel Maria Marian, Marco Garavelli, and Oliver Weingart. 2018. "QM/MM Photodynamics of Retinal in the Channelrhodopsin Chimera C1C2 with OM3/MRCI." *Chem Photo Chem*, November. doi:10.1002/cptc.201800185.

Dugave, Christophe. 2006. *Cis-Trans Isomerization in Biochemistry. Cis-Trans Isomerization in Biochemistry*. doi:10.1002/9783527609338.

Dugave, Christophe, and Luc Demange. 2003. "Cis-Trans Isomerization of Organic Molecules and Biomolecules: Implications and Applications." *Chemical Reviews*. doi:10.1021/cr0104375.

Dunkelberger, Adam D., Ryan D. Kieda, Jae Yoon Shin, Riccardo Rossi Paccani, Stefania Fusi, Massimo Olivucci, and F. Fleming Crim. 2012. "Photoisomerization and Relaxation Dynamics of a Structurally Modified Biomimetic Photoswitch." *Journal of Physical Chemistry A* 116 (14): 3527–33. doi:10.1021/jp300153a.

Erbas-Cakmak, Sundus, David A. Leigh, Charlie T. McTernan, and Alina L. Nussbaumer. 2015. "Artificial Molecular Machines." *Chemical Reviews*. doi:10.1021/acs.chemrev.5b00146.

Erdélyi, Máté, Anders Karlén, and Adolf Gogoll. 2005. "A New Tool in Peptide Engineering: A Photoswitchable Stilbene-Type β-Hairpin

Mimetic." *Chemistry - A European Journal* 12 (2): 403–12. doi:10.1002/chem.200500648.

Erdélyi, Máté, Miranda Varedian, Christian Sköld, Ida B. Niklasson, Johanna Nurbo, Åsa Persson, Jonas Bergquist, and Adolf Gogoll. 2008. "Chemistry and Folding of Photomodulable Peptides - Stilbene and Thioaurone-Type Candidates for Conformational Switches." *Organic and Biomolecular Chemistry* 6 (23): 4356–73. doi:10.1039/b812001c.

Ernst, Oliver P., David T. Lodowski, Marcus Elstner, Peter Hegemann, Leonid S. Brown, and Hideki Kandori. 2014. "Microbial and Animal Rhodopsins: Structures, Functions, and Molecular Mechanisms." *Chemical Reviews*. doi:10.1021/cr4003769.

Fang, Changfeng, Baswanth Oruganti, and Bo Durbeej. 2014. "Computational Study of the Working Mechanism and Rate Acceleration of Overcrowded Alkene-Based Light-Driven Rotary Molecular Motors." *RSC Advances* 4 (20): 10240–51. doi:10.1039/c3ra46880a.

Faulkner, Adele, Thomas Van Leeuwen, Ben L. Feringa, and Sander J. Wezenberg. 2016. "Allosteric Regulation of the Rotational Speed in a Light-Driven Molecular Motor." *Journal of the American Chemical Society* 138 (41): 13597–603. doi:10.1021/jacs.6b06467.

Fenno, Lief, Ofer Yizhar, and K Deisseroth. 2010. "The Development and Application of Optogenetics." *Annual Review of Neuroscience* 34 (1): 110301101035033. doi:10.1146/annurev-neuro-061010-113817.

Filatov, Michael, and Massimo Olivucci. 2014. "Designing Conical Intersections for Light-Driven Single Molecule Rotary Motors: From Precessional to Axial Motion." *Journal of Organic Chemistry* 79 (8): 3587–3600. doi:10.1021/jo5004289.

Frutos, L. M., T. Andruniow, F. Santoro, N. Ferre, and M. Olivucci. 2007. "Tracking the Excited-State Time Evolution of the Visual Pigment with Multiconfigurational Quantum Chemistry." *Proceedings of the National Academy of Sciences* 104 (19): 7764–69. doi:10.1073/pnas.0701732 104.

García-Iriepa, Cristina, Moussa Gueye, Jérémie Léonard, David Martínez-López, Pedro J. Campos, Luis Manuel Frutos, Diego Sampedro, and

Marco Marazzi. 2016. "A Biomimetic Molecular Switch at Work: Coupling Photoisomerization Dynamics to Peptide Structural Rearrangement." *Physical Chemistry Chemical Physics* 18 (9): 6742–53. doi:10.1039/C5CP07599H.

García-Iriepa, Cristina, Marco Marazzi, Luis Manuel Frutos, and Diego Sampedro. 2013. "E/Z Photochemical Switches: Syntheses, Properties and Applications." *RSC Advances* 3 (18). The Royal Society of Chemistry: 6241–66. doi:10.1039/c2ra22363e.

García-Iriepa, Cristina, Marco Marazzi, Felipe Zapata, Alessio Valentini, Diego Sampedro, and Luis Manuel Frutos. 2013. "Chiral Hydrogen Bond Environment Providing Unidirectional Rotation in Photoactive Molecular Motors." *Journal of Physical Chemistry Letters* 4 (9): 1389–96. doi:10.1021/jz302152v.

García-López, Víctor, Fang Chen, Lizanne G. Nilewski, Guillaume Duret, Amir Aliyan, Anatoly B. Kolomeisky, Jacob T. Robinson, Gufeng Wang, Robert Pal, and James M. Tour. 2017. "Molecular Machines Open Cell Membranes." *Nature* 548 (7669): 567–72. doi:10.1038/nature23657.

Gattuso, Hugo, Cristina García-Iriepa, Diego Sampedro, Antonio Monari, and Marco Marazzi. 2017. "Simulating the Electronic Circular Dichroism Spectra of Photoreversible Peptide Conformations." *Journal of Chemical Theory and Computation* 13 (7): 3290–96. doi:10.1021/acs.jctc.7b00163.

Gozem, Samer, Hoi Ling Luk, Igor Schapiro, and Massimo Olivucci. 2017. "Theory and Simulation of the Ultrafast Double-Bond Isomerization of Biological Chromophores." *Chemical Reviews* 117 (22): 13502–65. doi:10.1021/acs.chemrev.7b00177.

Graydon, Oliver. 2015. "Molecular Motors Driven by Light." *Nature Photonics* 9 (1). Nature Publishing Group: 13. doi:10.1038/nphoton.2014.312.

Guo, Yanan, Franziska E. Beyle, Beatrix M. Bold, Hiroshi C. Watanabe, Axel Koslowski, Walter Thiel, Peter Hegemann, Marco Marazzi, and Marcus Elstner. 2016. "Active Site Structure and Absorption Spectrum of Channelrhodopsin-2 Wild-Type and C128T Mutant." *Chemical Science* 7 (6): 3879–91. doi:10.1039/c6sc00468g.

Guo, Yanan, Franziska E. Wolff, Igor Schapiro, Marcus Elstner, and Marco Marazzi. 2018. "Different Hydrogen Bonding Environments of the Retinal Protonated Schiff Base Control the Photoisomerization in Channelrhodopsin-2." *Physical Chemistry Chemical Physics* DOI: 10.10. Royal Society of Chemistry. doi:10.1039/C8CP05210G.

Hasson, K C, F Gai, and P A Anfinrud. 1996. "The Photoisomerization of Retinal in Bacteriorhodospin: Experimental Evidence for a Three-State Model." *Proc. Natl. Acad. Sci. USA* 93 (26): 15124–29. doi:10.3934/dcdss.2018029.

Hegemann, Peter, and Andreas Möglich. 2011. "Channelrhodopsin Engineering and Exploration of New Optogenetic Tools." *Nature Methods* 8 (1): 39–42. doi:10.1038/nmeth.f.327.

Heyne, Karsten, Omar F. Mohammed, Anwar Usman, Jens Dreyer, Erik T. J. Nibbering, and Michael A. Cusanovich. 2005. "Structural Evolution of the Chromophore in the Primary Stages of Trans/Cis Isomerization in Photoactive Yellow Protein." *Journal of the American Chemical Society* 127 (51): 18100–106. doi:10.1021/ja051210k.

Hontani, Yusaku, Matthias Broser, Arita Silapetere, Benjamin S. Krause, Peter Hegemann, and John T. M. Kennis. 2017. "The Femtosecond-to-Second Photochemistry of Red-Shifted Fast-Closing Anion Channelrhodopsin PsACR1." *Physical Chemistry Chemical Physics* 19 (45): 30402–9. doi:10.1039/c7cp06414d.

Hontani, Yusaku, Marco Marazzi, Katja Stehfest, Tilo Mathes, Ivo H. M. Van Stokkum, Marcus Elstner, Peter Hegemann, and John T. M. Kennis. 2017. "Reaction Dynamics of the Chimeric Channelrhodopsin C1C2." *Scientific Reports* 7 (1). doi:10.1038/s41598-017-07363-w.

Hu, Deping, Jing Huang, Yu Xie, Ling Yue, Xuhui Zhuang, and Zhenggang Lan. 2015. "Nonadiabatic Dynamics and Photoisomerization of Biomimetic Photoswitches." *Chemical Physics* 463: 95–105. doi:10.1016/j.chemphys.2015.10.003.

J. Kastin, Abba, and Weihong Pan. 2010. "Concepts for Biologically Active Peptides." *Current Pharmaceutical Design* 16 (30): 3390–3400. doi:10.2174/138161210793563491.

Kandori, H., Y. Shichida, and T. Yoshizawa. 2001. "Photoisomerization in Rhodopsin." *Biochemistry (Moscow)*. doi:10.1023/A:1013123016803.

Kato, Hideaki E., Feng Zhang, Ofer Yizhar, Charu Ramakrishnan, Tomohiro Nishizawa, Kunio Hirata, Jumpei Ito, et al. 2012. "Crystal Structure of the Channelrhodopsin Light-Gated Cation Channel." *Nature* 482 (7385): 369–74. doi:10.1038/nature10870.

Kazaryan, Andranik, Jos C M J C M Jos C M Kistemaker, Lars V Schäfer, Wesley R W R Browne, Ben L B L Feringa, and Michael Filatov. 2010. "Understanding the Dynamics behind the Photoisomerization of a Light-Driven Fluorene Molecular Rotary Motor." *The Journal of Physical Chemistry. A* 114 (15): 5058–67. doi:10.1021/jp100609m.

Kazaryan, Andranik, Zhenggang Lan, Lars V. Schäfer, Walter Thiel, and Michael Filatov. 2011. "Surface Hopping Excited-State Dynamics Study of the Photoisomerization of a Light-Driven Fluorene Molecular Rotary Motor." *Journal of Chemical Theory and Computation* 7 (7): 2189–99. doi:10.1021/ct200199w.

Klare, Johann P., Igor Chizhov, and Martin Engelhard. 2008. "Microbial Rhodopsins: Scaffolds for Ion Pumps, Channels, and Sensors." *Results and Problems in Cell Differentiation*. doi:10.1007/400_2007_041.

Kort, R., H. Vonk, X. Xu, W. D. Hoff, W. Crielaard, and K. J. Hellingwerf. 1996. "Evidence for Trans-Cis Isomerization of the p-Coumaric Acid Chromophore as the Photochemical Basis of the Photocycle of Photoactive Yellow Protein." *FEBS Letters* 382 (1–2): 73–78. doi:10.1016/0014-5793(96)00149-4.

Koumura, Nagatoshi, Robert W J Zijistra, Richard A. Van Delden, Nobuyuki Harada, and Ben L. Feringa. 1999. "Light-Driven Monodirectional Molecular Rotor." *Nature* 401 (6749): 152–55. doi:10.1038/43646.

Lanyi, Janos K. 2004. "Bacteriorhodopsin." *Annual Review of Physiology* 66: 665–88. doi:10.1146/annurev.physiol.66.032102.150049.

Leeuwen, Thomas Van, Jasper Pol, Diederik Roke, Sander J. Wezenberg, and Ben L. Feringa. 2017. "Visible-Light Excitation of a Molecular Motor with an Extended Aromatic Core." *Organic Letters* 19 (6): 1402–5. doi:10.1021/acs.orglett.7b00317.

Liu, R. S., and A. E. Asato. 1985. "The Primary Process of Vision and the Structure of Bathorhodopsin: A Mechanism for Photoisomerization of Polyenes." *Proceedings of the National Academy of Sciences* 82 (2): 259–63. doi:10.1073/pnas.82.2.259.

Liu, R. S. H. 2001. "Photoisomerization by Hula-Twist: A Fundamental Supramolecular Photochemical Reaction." *Accounts of Chemical Research* 34 (7): 555–62. doi:10.1021/ar000165c.

London, Gabor, Kuang Yen Chen, Gregory T. Carroll, and Ben L. Feringa. 2013. "Towards Dynamic Control of Wettability by Using Functionalized Altitudinal Molecular Motors on Solid Surfaces." *Chemistry - A European Journal* 19 (32): 10690–97. doi:10.1002/chem.201300500.

Luk, Hoi Ling, Federico Melaccio, Silvia Rinaldi, Samer Gozem, and Massimo Olivucci. 2015. "Molecular Bases for the Selection of the Chromophore of Animal Rhodopsins." *Proceedings of the National Academy of Sciences* 112 (50): 15297–302. doi:10.1073/pnas.1510262112.

Lumento, Flavio, Vinicio Zanirato, Stefania Fusi, Elena Busi, Loredana Latterini, Fausto Elisei, Adalgisa Sinicropi, et al. 2007. "Quantum Chemical Modeling and Preparation of a Biomimetic Photochemical Switch." *Angewandte Chemie - International Edition* 46 (3): 414–20. doi:10.1002/anie.200602915.

Marchand, Gabriel, Julien Eng, Igor Schapiro, Alessio Valentini, Luis Manuel Frutos, Elisa Pieri, Massimo Olivucci, Jérémie Léonard, and Etienne Gindensperger. 2015. "Directionality of Double-Bond Photoisomerization Dynamics Induced by a Single Stereogenic Center." *Journal of Physical Chemistry Letters* 6 (4): 599–604. doi:10.1021/jz502644h.

Mathies, RA, and J Lugtenburg. 2000. *Handbook of Biological Physics*. Edited by Pugh Stavenga, DG, de Grip, WJ. Vol. 3. Elsevier Science B. V.

Matsuno-Yagi, Akemi, and Yasuo Mukohata. 1977. "Two Possible Roles of Bacteriorhodopsin; a Comparative Study of Strains of Halobacterium Halobium Differing in Pigmentation." *Biochemical and Biophysical Research Communications* 78 (1): 237–43. doi:10.1016/0006- 291X (77)91245-1.

Melaccio, Federico, Nicolas Calimet, Igor Schapiro, Alessio Valentini, Marco Cecchini, and Massimo Olivucci. 2016. "Space and Time Evolution of the Electrostatic Potential during the Activation of a Visual Pigment." *Journal of Physical Chemistry Letters* 7 (13): 2563–67. doi:10.1021/acs.jpclett.6b00977.

Melloni, Alfonso, Riccardo Rossi Paccani, Donato Donati, Vinicio Zanirato, Adalgisa Sinicropi, Maria Laura Parisi, Elena Martin, et al. 2010. "Modeling, Preparation, and Characterization of a Dipole Moment Switch Driven by Z/E Photoisomerization." *Journal of the American Chemical Society* 132 (27): 9310–19. doi:10.1021/ja906733q.

Nagel, Georg, Doris Ollig, Markus Fuhrmann, Suneel Kateriya, Anna Maria Musti, Ernst Bamberg, and Peter Hegemann. 2002. "Channelrhodopsin-1: A Light-Gated Proton Channel in Green Algae." *Science* 296 (5577): 2395–98. doi:10.1126/science.1072068.

Nakamichi, Hitoshi, Volker Buss, and Tetsuji Okada. 2007. "Photoisomerization Mechanism of Rhodopsin and 9-Cis-Rhodopsin Revealed by X-Ray Crystallography." *Biophysical Journal* 92 (12). doi:10.1529/biophysj.107.108225.

Neutze, Richard, Eva Pebay-Peyroula, Karl Edman, Antoine Royant, Javier Navarro, and Ehud M. Landau. 2002. "Bacteriorhodopsin: A High-Resolution Structural View of Vectorial Proton Transport." *Biochimica et Biophysica Acta - Biomembranes*. doi:10.1016/S0005-2736(02) 00566-7.

Nomura, Yujiro, Takashi Bando, Yoshito Takeuchi, and Shuji Tomoda. 1983. "Synthesis of Cyclic Imines Having Conjugated Exocyclic Double Bond." *Bulletin of the Chemical Society of Japan* 56 (10). The Chemical Society of Japan: 3199–3200. doi:10.1246/bcsj.56.3199.

Oesterhelt, Dieter, and Walther Stoeckenius. 1971. "Rhodopsin-like Protein from the Purple Membrane of Halobacterium Halobium." *Nature New Biology* 233 (39): 149–52. doi:10.1038/newbio233149a0.

Oruganti, Baswanth, Changfeng Fang, and Bo Durbeej. 2015. "Computational Design of Faster Rotating Second-Generation Light-Driven Molecular Motors by Control of Steric Effects." *Physical Chemistry Chemical Physics* 17 (33): 21740–51. doi:10.1039/c5cp02303c.

Oruganti, Baswanth, Jun Wang, and Bo Durbeej. 2016. "Computational Insight to Improve the Thermal Isomerisation Performance of Overcrowded Alkene-Based Molecular Motors through Structural Redesign." *Chem Phys Chem* 17 (21): 3399–3408. doi:10.1002/cphc.201600766.

Pérez-Hernández, Guillermo, and Leticia González. 2010. "Mechanistic Insight into Light-Driven Molecular Rotors: A Conformational Search in Chiral Overcrowded Alkenes by a Pseudo-Random Approach." *Physical Chemistry Chemical Physics* 12 (38): 12279–89. doi:10.1039/c0cp00324g.

Polli, Dario, Piero Altoè, Oliver Weingart, Katelyn M. Spillane, Cristian Manzoni, Daniele Brida, Gaia Tomasello, et al. 2010. "Conical Intersection Dynamics of the Primary Photoisomerization Event in Vision." *Nature* 467 (7314): 440–43. doi:10.1038/nature09346.

Quail, Peter H., Margaret T. Boylan, Brian M. Parks, Timothy W. Short, Yong Xu, and Doris Wagner. 1995. "Phytochromes: Photosensory Perception and Signal Transduction." *Science* 268 (5211): 675–80. doi:10.1126/science.7732376.

Rafiq, Shahnawaz, Basanta Kumar Rajbongshi, Nisanth N Nair, Pratik Sen, and Gurunath Ramanathan. 2011. "Excited State Relaxation Dynamics of Model Green Fluorescent Protein Chromophore Analogs : Evidence for Cis - Trans Isomerism." *Journal of Physical Chemistry A* 115 (47): 13733–42. doi:10.1021/jp206815t.

Regner, N., T. T. Herzog, K. Haiser, C. Hoppmann, M. Beyermann, J. Sauermann, M. Engelhard, T. Cordes, K. Rück-Braun, and W. Zinth. 2012. "Light-Switchable Hemithioindigo-Hemistilbene-Containing Peptides: Ultrafast Spectroscopy of the Z → e Isomerization of the Chromophore and the Structural Dynamics of the Peptide Moiety." *Journal of Physical Chemistry B* 116 (14): 4181–91. doi:10.1021/jp300982a.

Renner, Christian, and Luis Moroder. 2006. "Azobenzene as Conformational Switch in Model Peptides." *ChemBioChem.* doi:10.1002/cbic.200500531.

Ritter, Eglof, Katja Stehfest, Andre Berndt, Peter Hegemann, and Franz J. Bartl. 2008. "Monitoring Light-Induced Structural Changes of Channelrhodopsin-2 by UV-Visible and Fourier Transform Infrared Spectroscopy." *Journal of Biological Chemistry* 283 (50): 35033–41. doi:10.1074/jbc.M806353200.

Rivado-Casas, Laura, Marina Blanco-Lomas, Pedro J. Campos, and Diego Sampedro. 2011. "Photochemical Characterization of Biomimetic Molecular Switches." *Tetrahedron* 67 (39): 7570–74. doi:10.1016/j.tet.2011.07.070.

Rivado-Casas, Laura, Pedro J. Campos, and Diego Sampedro. 2010. "Reaction of Fischer Alkynyl Carbene Complexes with Fluorenone Imines: Mechanistic Studies." *Organometallics* 29 (14): 3117–24. doi:10.1021/om100219h.

Rivado-Casas, Laura, Diego Sampedro, Pedro J. Campos, Stefania Fusi, Vinicio Zanirato, and Massimo Olivucci. 2009. "Fluorenylidene-Pyrroline Biomimetic Light-Driven Molecular Switches." *Journal of Organic Chemistry* 74 (13): 4666–74. doi:10.1021/jo802792j.

Rohmer, Thierry, Christina Lang, Christian Bongards, Karthick Babu Sai Sankar Gupta, Johannes Neugebauer, Jon Hughes, Wolfgang Gärtner, and Jörg Matysik. 2010. "Phytochrome as Molecular Machine: Revealing Chromophore Action during the Pfr → Pr Photoconversion by Magic-Angle Spinning NMR Spectroscopy." *Journal of the American Chemical Society* 132 (12): 4431–37. doi:10.1021/ja9108616.

Russew, Maria-Melanie, and Stefan Hecht. 2010. "Photoswitches: From Molecules to Materials." *Advanced Materials* 22 (31). WILEY-VCH Verlag: 3348–60. doi:10.1002/adma.200904102.

Sadovski, Oleg, Andrew A. Beharry, Fuzhong Zhang, and G. Andrew Woolley. 2009. "Spectral Tuning of Azobenzene Photoswitches for Biological Applications." *Angewandte Chemie - International Edition* 48 (8): 1484–86. doi:10.1002/anie.200805013.

Sampedro, Diego, Annapaola Migani, Alessandra Pepi, Elena Busi, Riccardo Basosi, Loredana Latterini, Fausto Elisei, et al. 2004. "Design and Photochemical Characterization of a Biomimetic Light-Driven Z/E Switcher." *Journal of the American Chemical Society* 126 (30): 9349–59. doi:10.1021/ja038859e.

Sasaki, Jun, and John L. Spudich. 2008. "Signal Transfer in Haloarchaeal Sensory Rhodopsin-Transducer Complexes." In *Photochemistry and Photobiology*, 84:863–68. doi:10.1111/j.1751-1097.2008.00314.x.

Saywell, Alex, Anne Bakker, Johannes Mielke, Takashi Kumagai, Martin Wolf, Víctor García-López, Pinn Tsong Chiang, James M. Tour, and Leonhard Grill. 2016. "Light-Induced Translation of Motorized Molecules on a Surface." *ACS Nano* 10 (12): 10945–52. doi:10.1021/acsnano.6b05650.

Schapiro, Igor, and Sanford Ruhman. 2014. "Ultrafast Photochemistry of Anabaena Sensory Rhodopsin: Experiment and Theory." *Biochimica et Biophysica Acta - Bioenergetics*. doi:10.1016/j.bbabio.2013.09.014.

Schapiro, Igor, Mikhail Nikolaevich Ryazantsev, Luis Manuel Frutos, Nicolas Ferré, Roland Lindh, and Massimo Olivucci. 2011. "The Ultrafast Photoisomerizations of Rhodopsin and Bathorhodopsin Are Modulated by Bond Length Alternation and HOOP Driven Electronic Effects." *Journal of the American Chemical Society* 133 (10): 3354–64. doi:10.1021/ja1056196.

Schierling, B., A.-J. Noel, W. Wende, L. T. Hien, E. Volkov, E. Kubareva, T. Oretskaya, et al. 2010. "Controlling the Enzymatic Activity of a Restriction Enzyme by Light." *Proceedings of the National Academy of Sciences* 107 (4): 1361–66. doi:10.1073/pnas.0909444107.

Schoenlein, R. W., L. A. Peteanu, R. A. Mathies, and C. V. Shank. 1991. "The First Step in Vision: Femtosecond Isomerization of Rhodopsin." *Science* 254 (5030): 412–15. doi:10.1126/science.1925597.

Shibata, Mikihiro, Taro Tanimoto, and Hideki Kandori. 2003. "Water Molecules in the Schiff Base Region of Bacteriorhodopsin." *Journal of the American Chemical Society* 125 (44): 13312–13. doi:10.1021/ja037343s.

Siampiringue, Narcisse, Ghislain Guyot, Sandra Monti, and Pietro Bortolus. 1987. "The Cis → Trans Photoisomerization of Azobenzene: An Experimental Re-Examination." *Journal of Photochemistry* 37 (1): 185–88. doi:10.1016/0047-2670(87)85039-6.

Sineshchekov, O. A., K.-H. Jung, and J. L. Spudich. 2002. "Two Rhodopsins Mediate Phototaxis to Low- and High-Intensity Light in Chlamydomonas Reinhardtii." *Proceedings of the National Academy of Sciences* 99 (13): 8689–94. doi:10.1073/pnas.122243399.

Sinicropi, A., E. Martin, M. Ryazantsev, J. Helbing, J. Briand, D. Sharma, J. Leonard, et al. 2008. "An Artificial Molecular Switch That Mimics the Visual Pigment and Completes Its Photocycle in Picoseconds." *Proceedings of the National Academy of Sciences* 105 (46): 17642–47. doi:10.1073/pnas.0802376105.

Sinicropi, Adalgisa, Caterina Bernini, Riccardo Basosi, and Massimo Olivucci. 2009. "A Novel Biomimetic Photochemical Switch at Work: Design of a Photomodulable Peptide." *Photochemical and Photobiological Sciences*. doi:10.1039/b906271h.

Slamovits, Claudio H., Noriko Okamoto, Lena Burri, Erick R. James, and Patrick J. Keeling. 2011. "A Bacterial Proteorhodopsin Proton Pump in Marine Eukaryotes." *Nature Communications* 2 (1). doi:10.1038/ncomms1188.

Sluysmans, Damien, and J. Fraser Stoddart. 2018. "Growing Community of Artificial Molecular Machinists." *Proceedings of the National Academy of Sciences* 115 (38): 9359–61. doi:10.1073/pnas.1813973115.

Smitienko, O a, M N Mozgovaya, I V Shelaev, F E Gostev, T B Feldman, V a Nadtochenko, O M Sarkisov, and M a Ostrovsky. 2010. "Femtosecond Formation Dynamics of Primary Photoproducts of Visual Pigment Rhodopsin." *Biochemistry. Biokhimii͡a* 75 (1): 25–35. doi:10.1134/S0006297910010049.

Strambi, A., B. Durbeej, N. Ferre, and M. Olivucci. 2010. "Anabaena Sensory Rhodopsin Is a Light-Driven Unidirectional Rotor." *Proceedings of the National Academy of Sciences* 107 (50): 21322–26. doi:10.1073/pnas.1015085107.

Szymański, Wiktor, John M. Beierle, Hans A.V. Kistemaker, Willem A. Velema, and Ben L. Feringa. 2013. "Reversible Photocontrol of Biological Systems by the Incorporation of Molecular Photoswitches." *Chemical Reviews*. doi:10.1021/cr300179f.

Vachon, Jérôme, Gregory T. Carroll, Michael M. Pollard, Emile M. Mes, Albert M. Brouwer, and Ben L. Feringa. 2014. "An Ultrafast Surface-Bound Photo-Active Molecular Motor." *Photochemical and Photobiological Sciences* 13 (2): 241–46. doi:10.1039/c3pp50208b.

Voliani, Valerio, Ranieri Bizzarri, Riccardo Nifos, Stefania Abbruzzetti, Elena Grandi, Cristiano Viappiani, and Fabio Beltram. 2008. "Cis-Trans Photoisomerization of Fluorescent-Protein Chromophores." *Journal of Physical Chemistry B* 112 (34): 10714–22. doi:10.1021/jp802419h.

Volkov, Oleksandr, Kirill Kovalev, Vitaly Polovinkin, Valentin Borshchevskiy, Christian Bamann, Roman Astashkin, Egor Marin, et al. 2017. "Structural Insights into Ion Conduction by Channelrhodopsin 2." *Science* 358 (6366). doi:10.1126/science.aan8862.

Wang, Jun, and Bo Durbeej. 2018. "Toward Fast and Efficient Visible-Light-Driven Molecular Motors: A Minimal Design." *Chemistry Open* 7 (8): 583–89. doi:10.1002/open.201800089.

Wang, Jun, Baswanth Oruganti, and Bo Durbeej. 2017. "Light-Driven Rotary Molecular Motors without Point Chirality: A Minimal Design."

Physical Chemistry Chemical Physics 19 (10): 6952–56. doi:10.1039/c6cp08484b.

Warshel, A., Z. T. Chu, and J. K. Hwang. 1991. "The Dynamics of the Primary Event in Rhodopsins Revisited." *Chemical Physics* 158 (2–3): 303–14. doi:10.1016/0301-0104(91)87074-6.

Warshel, Arieh. 1976. "Bicycle-Pedal Model for the First Step in the Vision Process." *Nature* 260 (5553): 679–83. doi:10.1038/260679a0.

Warshel, Arieh, and Natalia Barboy. 1982. "Energy Storage and Reaction Pathways in the First Step of the Vision Process." *Journal of the American Chemical Society* 104 (6): 1469–76. doi:10.1021/ja00370a003.

Watanabe, Hiroshi C., Kai Welke, Franziska Schneider, Satoshi Tsunoda, Feng Zhang, Karl Deisseroth, Peter Hegemann, and Marcus Elstner. 2012. "Structural Model of Channelrhodopsin." *Journal of Biological Chemistry* 287 (10): 7456–66. doi:10.1074/jbc.M111.320309.

Welke, Kai, Jan S. Frähmcke, Hiroshi C. Watanabe, Peter Hegemann, and Marcus Elstner. 2011. "Color Tuning in Binding Pocket Models of the Chlamydomonas-Type Channelrhodopsins." *Journal of Physical Chemistry B* 115 (50): 15119–28. doi:10.1021/jp2085457.

Wezenberg, Sander J., Kuang Yen Chen, and Ben L. Feringa. 2015. "Visible-Light-Driven Photoisomerization and Increased Rotation Speed of a Molecular Motor Acting as a Ligand in a Ruthenium(II) Complex." *Angewandte Chemie - International Edition* 54 (39): 11457–61. doi:10.1002/anie.201505781.

Yizhar, Ofer, Lief E. Fenno, Thomas J. Davidson, Murtaza Mogri, and Karl Deisseroth. 2011. "Optogenetics in Neural Systems." *Neuron*. doi:10.1016/j.neuron.2011.06.004.

Yu, Zhilin, and Stefan Hecht. 2016. "Remote Control over Folding by Light." *Chemical Communications* 52 (40): 6639–53. doi:10.1039/c6cc01423b.

Zanirato, Vinicio, Gian P. Pollini, Carmela De Risi, Filippo Valente, Alfonso Melloni, Stefania Fusi, Jacopo Barbetti, and Massimo Olivucci. 2007. "Synthesis of Biomimetic Light-Driven Molecular Switches via a Cyclopropyl Ring-Opening/Nitrilium Ion Ring-Closing Tandem

Reaction." *Tetrahedron* 63 (23): 4975–82. doi:10.1016/ j.tet.2007.03.141.

Zhao, Depeng, Thomas Van Leeuwen, Jinling Cheng, and Ben L. Feringa. 2017. "Dynamic Control of Chirality and Self-Assembly of Double-Stranded Helicates with Light." *Nature Chemistry* 9 (3): 250–56. doi:10.1038/nchem.2668.

In: Photoisomerization
Editor: Diego Sampedro
ISBN: 978-1-53615-313-2

Chapter 6

PHOTOISOMERIZATION OF CARBON-CARBON DOUBLE BONDS AS KEY STEP IN ORGANIC SYNTHESIS

Javier I. Bardagi, Sofía Caby, Silvia M. Soria-Castro and Gabriela Oksdath-Mansilla*
INFIQC-CONICET-UNC, Dpto. de Química Orgánica,
Facultad de Ciencias Químicas, Universidad Nacional de Córdoba,
Ciudad Universitaria, Córdoba, Argentina

ABSTRACT

Synthetic organic photochemistry constitutes an active research area and powerful tool for the preparation of natural products as well as others molecules with high structural complexity, in a simple way under mild conditions. In this context, numerous light promoted transformations are reported every year demonstrating the versatility and scope of the photochemical reactions.

In particular, photoisomerization of double bonds has been extensively studied evaluating the conformational or geometrical changes induced on

* Corresponding Author's Email: goksdath@fcq.unc.edu.ar.

the molecule. However, spatiotemporal control of specific alkene geometry has attractive interest in organic synthesis, especially when the success of a desired transformation depend on a predefined configuration. There are many synthetic strategies which involve the photoisomerization as a key step to access to new molecular structures, proving the synthetic potential of this reaction.

Therefore, in this chapter we focus on the synthetic point of view of photoisomerization as efficient method to facilitate a spatiotemporal control of an alkene, at a defined stage of a synthetic sequence. The synthetic aspects of light-induced isomerization will be described, evaluating its participation in combination with other reactions such as cyclization, Pd-catalysis, radical addition, hydroamination or Stetter reaction. Using an appropriate synthetic strategy, valuable derivatives such as phenanthridinones, quinolinones, cyclic and spirocyclic ketals, chromenes or coumarin can be obtained. Furthermore, photoisomerization applied to the synthesis of natural product or molecules with biological relevance are discussed.

INTRODUCTION

Inspired by the nature, synthetic organic chemists are continuously developing new chemical protocols, proposing new challenges regarding the state of the art on synthetic methodology. In order to approach to "ideal synthesis," the processes should be focus in providing large quantities of complex products, with an energy-efficient methodology and minimizing work, materials expenses and waste. In this sense, photochemical transformations are excellent tools to be considered for the synthesis of diverse organic frameworks containing different functional groups. Considering that the electronically excited molecules show different chemical reactivity in regard to its ground state, reactions can be achieved from excited molecules that otherwise are unviable or difficult to carry out. Photochemical transformations have been successfully applied as key steps for creating both natural products and molecules with high structural complexity, in a simple way and under mild conditions [1, 2].

In this context, light-induced reactions of alkenes could be of great synthetic interest including, inter- and intramolecular cycloadditions, photooxidations, electrocyclic ring opening and closing of conjugated dienes and polyenes, and *E-Z* isomerization [3].

In nature, many biological processes depend of photoisomerization reactions allowing a regulation of the activity of biomolecules through a temporal and spatial conformational control. For instance, the light-promoted isomerization of (*Z*)-retinal generates conformational changes in the protein, resulting in a variety of biological functions such as vision, photosensing, and ion transportation [4]. Another interesting process is the biosynthesis of coumarin which involves the participation of 4'-coumaroyl-*S*-Coenzyme A as a crucial intermediate. The *E* → *Z* isomerization of the exocyclic double bond in this adduct is crucial to carry out the final lactonization reaction [5].

Double-bond isomerization focused on geometrical control is probably one of the most prominent photochemical transformations of alkenes. In this case, the isomerization is governed by the ratio of the concentration of *E-Z* isomers, the corresponding quantum yields (Φ_{E-Z} and Φ_{Z-E}, respectively) and the molar absorption coefficients. On the other hand, considering that the isomerization of alkenes involve the excitation of an electron from a π orbital to a π* orbital, it can occur via singlet (S_1) or triplet (T_1) excited states, leading to a twisted intermediate with zwitterionic or diradical character for S_1 or T_1 respectively.

For isolated double bound the intersystem crossing (ISC) between S_1 and T_1 tends to be inefficient, therefore the isomerization by direct excitation occurs from singlet state. With the presence of conjugated π system the ISC efficiency is increased and a triplet state can be involved in photoisomerization. Taking into account the photophysic properties of the alkenes, the isomerization reaction can be achieved selectively and directionally by excitation at a specific wavelength of one of the isomers. Photosensitization is also possible using a photosensitizer which possess excited-state energy between the energy of excited (*E*)- and (*Z*)-isomer [6].

Photoisomerization has been successfully applied to organic synthesis and nowadays is consider a key step in many synthetic methodologies, accessing molecules with structural complexity that are relevant in areas such as medicine or biology.

In this chapter we highlight the synthetic utility of the photo-isomerization to obtain molecules with relevance in organic chemistry. We focus on the *E-Z* isomerization under direct or sensitized excitation, as a key step in a sequence of reactions towards a target molecule synthesis. In a source of organization, we started with direct and sensitized photoisomerization in rather short two-step synthesis, showing the potential of photoisomerization in sequential and one-pot transformation. Next, special cases of double bond isomerization in cyclic system were addressed to finally show the utility of the approach in complex natural product synthesis.

ALKENE PHOTOISOMERIZATIONS

Direct Excitation

As part of a study for the synthesis of substituted indazoles under mild, base promoted, transition-metal-free conditions, Bolm and co-workers used a UV driven photoisomerization of *N*-tosylhydrazone [7]. After a conscious optimization of the cyclization reaction of (Z)-2-bromoacetophenone *N*-tosylhydrazones (Scheme 1), the authors realized that the (*E*)-isomer was totally unreactive. From the analysis of the total synthetic approach, it was recognized that the synthesis of pure (*Z*)-isomer was a weakness of the method so an approach that could use a mixture of *Z*:*E* isomers would make the strategy more attractive. An improved yield relate to the "dark" reaction was obtained performing a reaction of *Z-E* mixtures of 2-bromoacetophenone *N*-tosylhydrazones with UV light irradiation but in a poor extent (Scheme 1). However, if more optimization of the time irradiated reaction would done, probably better result could be obtained.

A great potential of photoisomerization resides in the possibility to perform a known transformation in milder condition, providing a more reactive conformer based in two reason: energy and "right geometry". The acid catalyzed synthesis of ketal and spiroketals from (*Z*)-δ-hydroxy enones development by Smith et al. is an excellent example of this advantage [8].

The author recognized that the harsh conditions needed (strongly acidic) to transform (*Z*)-*δ*-hydroxy enones to cyclic ketal preclude its use in natural product synthesis, despite of the excellent dr excess of the method. The proposal of a photoisomerization - catalytic acid cyclization proved to be an optimum mild condition to ketals and spiroketals with a very good dr excess (Scheme 2). The proposed mechanism was supported by sequential reactions showing the (*Z*)-isomer to be unreactive to catalytic acid cyclization.

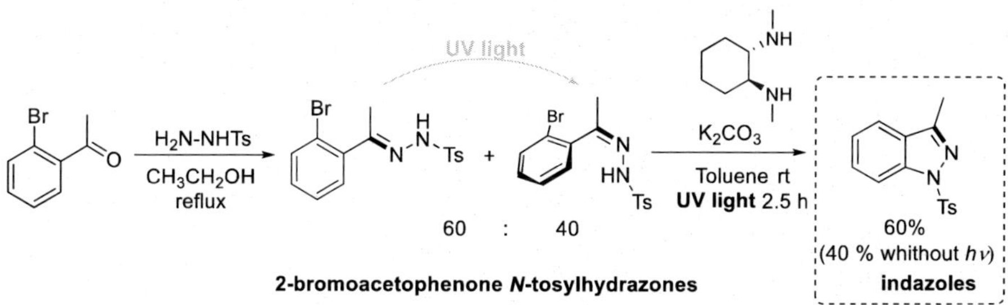

Scheme 1. Synthesis of indazoles from 2-bromoacetophenones.

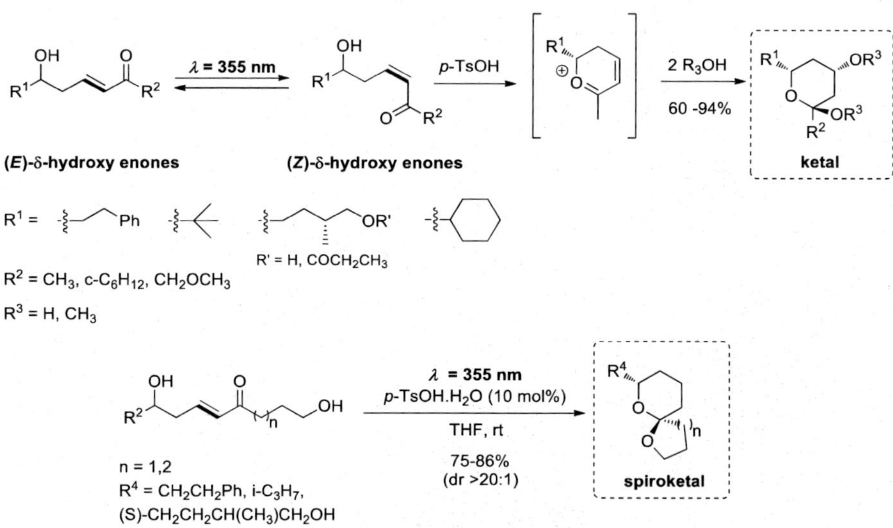

Scheme 2. Acid catalyzed synthesis of ketals.

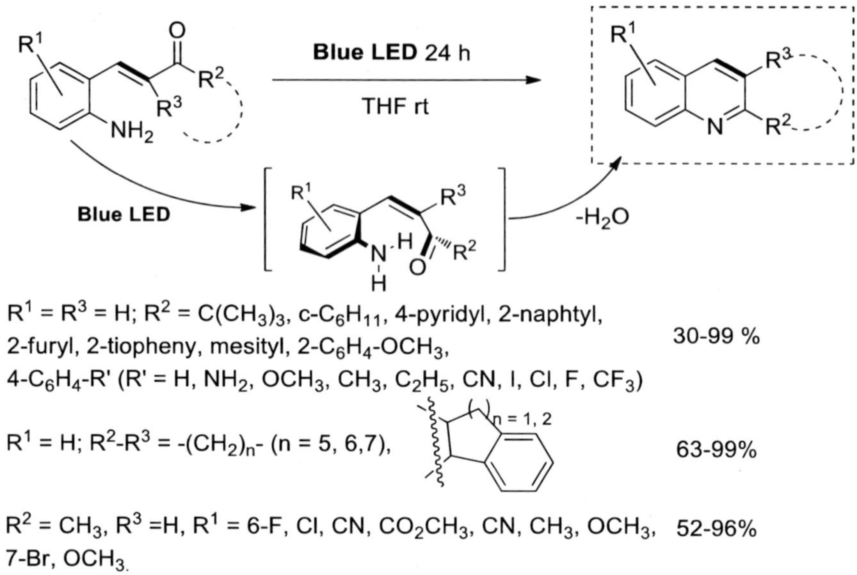

Scheme 3. Condensation of 2-(2-aminoaryl)-1,2-unsaturated ketones to quinolones.

Another example of photoisomerization – cyclization reaction was developed to obtain substituted quinolines from α,β-unsaturated ketones (Scheme 3) [9]. Direct visible-light (blue LEDs) irradiation driven a spontaneous condensations of the (*Z*)-isomer photoproduct with excellent yields. Exploration of the scope showed a good tolerance to ketone and aryl substitution being an extra amino group in the substrate the only exception.

Coupling of photoisomerization with cycloaddition reactions could be a good approach to obtain heterocyclic compounds. In this context, Ishar and co-workers reported a synthetic methodology focused on the preparation of stable endoperoxides (Scheme 4) [10]. Following their work of 1989 [11, 12], they carried out the irradiation of (*E*,*E*)-arylidine-β-ionones in thiophene-free dry benzene with constant bubbling of oxygen. By photoisomerization and 6π-electrocyclization, the tetrahydrobenzopyran is formed in situ, which undergoes a regioselective photocycloaddition with molecular oxygen to give rise to the corresponding endoperoxide. The arylidene-β-ionone system has been extensively studied by the authors, who demonstrated that the photo-reactivity of these compounds depends markedly of the reaction conditions (solvent, atmosphere) and the electronic nature of the substituents of the substrate [13]. They observed that with electron rich arylidene-*β*-ionone, there was no formation of endoperoxides. However, if the reaction is performed under N_2 atmosphere, the intermediate tetrahydrobenzopyran are obtained, which upon further irradiation in the presence of oxygen afforded endoperoxides. This behavior was attributed to the readily decomposition by over oxidation of the endoperoxide product possessing electron rich aryl moieties, to give rise to lactone derivatives.

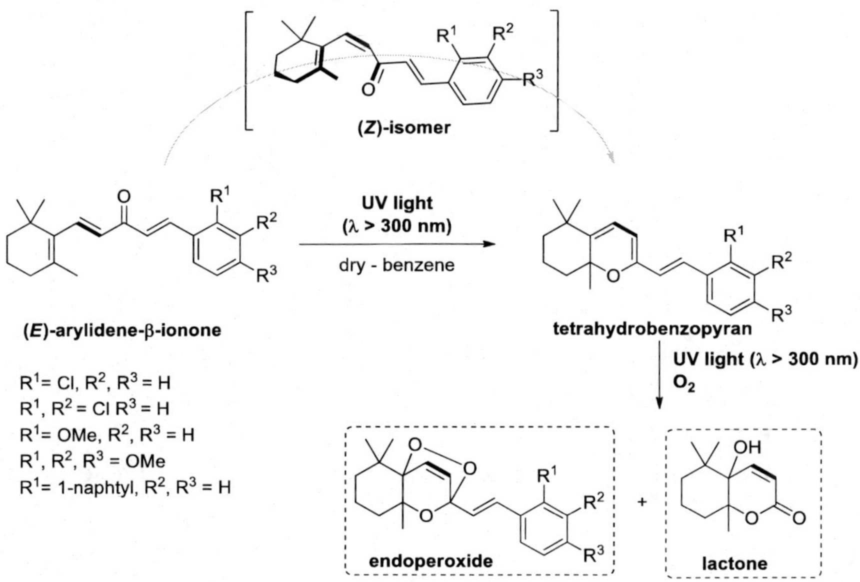

Scheme 4. Photoisomerization-electrocyclization to tetrahydrobenzopyrans and its photoreaction with molecular oxygen.

Sensitized Excitation

From the beginning of photoinduced transformation, energy transfer reaction (sensitization) has been recognized as a good approach. Sensitized reactions have several advantages related to direct irradiation, from which we can highlight the access to excited state of not absorbing substrates at the working wavelength. This feature usually guides us to better selectivity and broad applicability of the method.

The simplest approach of a photoisomerization couple to a following chemical reaction could be exemplified in the photoisomerization-spontaneous cyclization(polar) reaction of ortho-hydroxycinnamates [14]. Zhan and Li showed in this work the synthesis of coumarin by a sensitized isomerization of (*E*)-ortho-hydroxycinnamates induced by an Ir catalyst and blue light followed by the spontaneous cyclization (Scheme 5). The authors based their work in the direct photoisomerization of the same substrates development by Suzuki et al. in 2002 [15]. The results of this earlier (more detailed) work described that cyclization of (*Z*)-ortho-hydroxycinnamates take place during work-up, possible by heating, and not during the irradiation time. This seam also be the case in the Zhan and Li's work but it was not specified by the authors.

Gilmour's group synthesized CF_3 substituted *2H*-chromenes working with cinnamic alcohol using a different strategy: a two-step synthesis composed of a photoisomerization followed by a Pd catatalyzed C-O formation [16]. In this work, sensitized isomerization was first study due to known photodegradation of cinnamic alcohols. The authors conclude that CF_3 group at 4 position has a big influence in the photo-stability allowing the quantitative transformation of $Z \rightarrow E$ using anthracene as photocatalyst (Scheme 6).

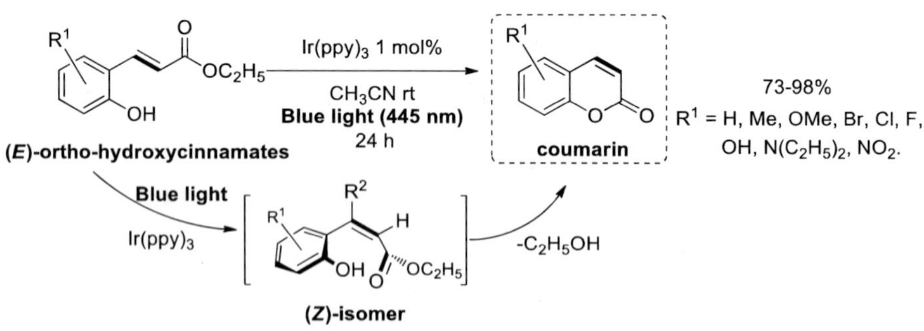

Scheme 5. Sensitized synthesis of coumarin.

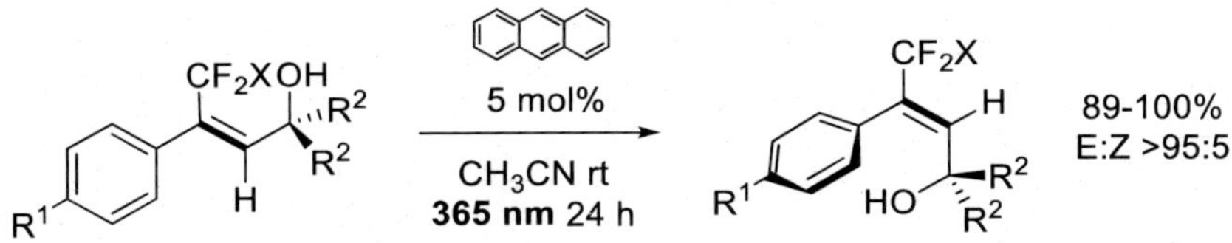

R^1 = H, OCH_3, CH_3, Cl, F.
R^2 = H, CH_3.
X = H, F.

Scheme 6. $Z \rightarrow E$ sensitized isomerization of CF_3 substituted cinnamic alcohols.

For the Pd catalyzed cyclization an ortho leaving group was necessary in the aryl group of *Z* cinnamic alcohols substrates. Inclusion of bromide in this position did not affect photo-isomerization reaction allowing the formation of the *E* isomer in good yields (Scheme 7). Finally, Pd catalyzed reaction was accomplished by a reported procedure [17] for the synthesis of 4-aryl-2*H*-chromenes without further optimization. It is important to note that isomerization step could not be done in the Pd catalyzed reaction conditions, even at high temperature, as it was demonstrated by the reaction of the *Z*-configured allylic alcohols.

The powerful of this kind of photo-catalyzed isomerization – metal catalyzed sequence was also shown by Gilmour and Watson in an intermolecular version [18]. In this work the sensitized *E* to *Z* isomerization of important styrenyl boron species was accomplished with an Ir catalyst in good to excellent yields (Scheme 8).

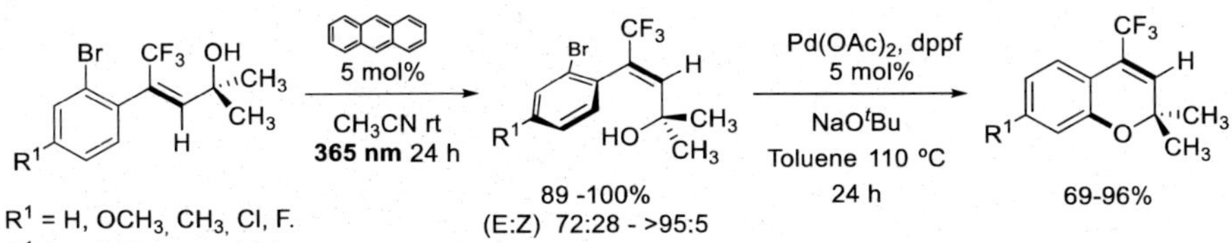

R^1 = H, OCH_3, CH_3, Cl, F.
R^1-Ar = 2-Bromonaphtyl.

Scheme 7. Synthesis of 2*H*-chromenes by a photoisomerization - Pd catatalyzed C-O bond formation sequence.

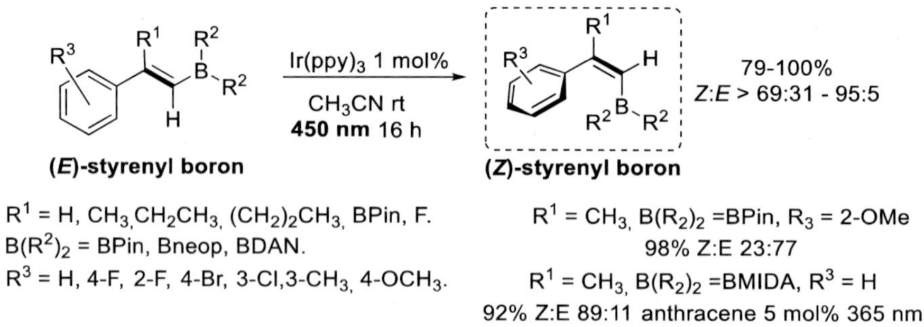

Scheme 8. Photo-catalyzed isomerization of styrenyl boron derivatives.

Although the scope of the following transformation of the (*Z*)-isomer in Suzuki coupling was not extensively explored, the author showed the potential of this strategy performing different transformation of the *Z* boron derivative with total conservation of the stereochemistry of the double bond, opening an efficient way to explore two-dimensional chemical space (Scheme 9).

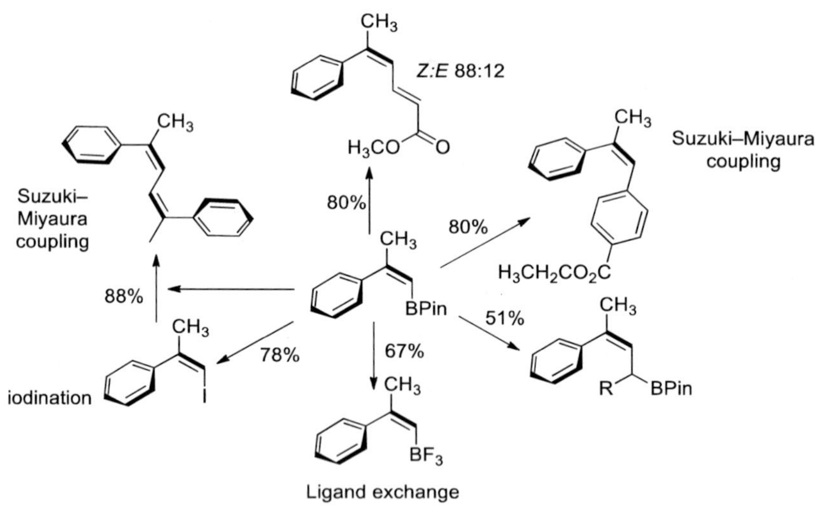

Scheme 9. Transformation of (*Z*)-styrenyl boron derivatives with conservation of double bond conformation.

Cognizant of the potential of cooperative chemoenzymatic reaction, Hartwig's and Zhao's groups development a photocatalyzed isomerization -

enzymatic reduction sequence of alkenes to prepare enantiopure products [19]. The authors used a flavin mononucleotide (FMN) and an Ir catalyst ($[Ir(dmppy)_2(dtbbpy)]PF_6$) for the sensitized isomerization of (*Z*)-phenylbut-2-enedioic acid dimethyl ester to the (*E*)-isomer, which was the substrate for an Ene-reductase (ER) (Scheme 10). It is important to point out that the isomerization reaction was optimized in the conditions in which the enzymatic reaction works, with the necessary co-factor, which was different of previously reported isomerization. Besides the transformation was not limited to diester derivatives with the reaction being tolerant to combinations of CN, C(O)R, C(O)NRR' and CF_3 groups (Scheme 10).

The selectivity of the enzymatic reaction for one of the isomers allows a high yield and *ee* even in the case of an unfavorable *Z*:*E* ratio, which highlight the great advantage of a cooperative catalysis in contrast to a sequential transformation (Scheme 11). With the cooperative isomerization and enzymatic reduction system, enantioenriched compounds can be obtained, which are precursors to valuable synthetic intermediates and have been used in the synthesis of biologically active compounds and natural products [20, 21, 22].

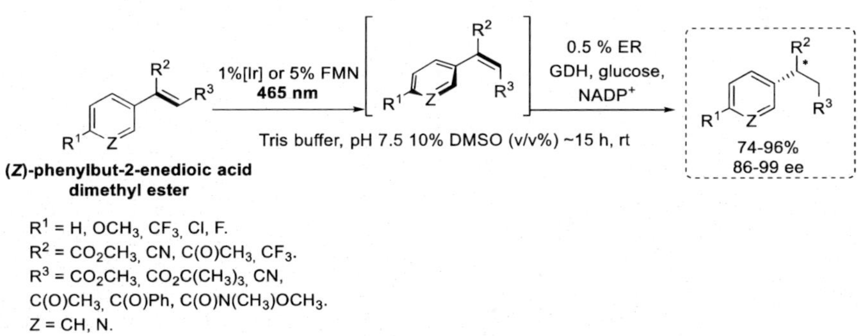

Scheme 10. Cooperative photocatalyzed isomerization - enzymatic reduction.

Another kind of cooperative photocatalyzed - photocatalyzed single catalysts transformation, driven by visible-light, was applied recently using cinnamic acids and amides [23, 24]. This elegant transformation allows an intramolecular cyclization by the substitution of two H atoms with a very good atom economy. This fact, together with the use of visible light, makes this procedure a very desirable transformation in the context of Green Chemistry.

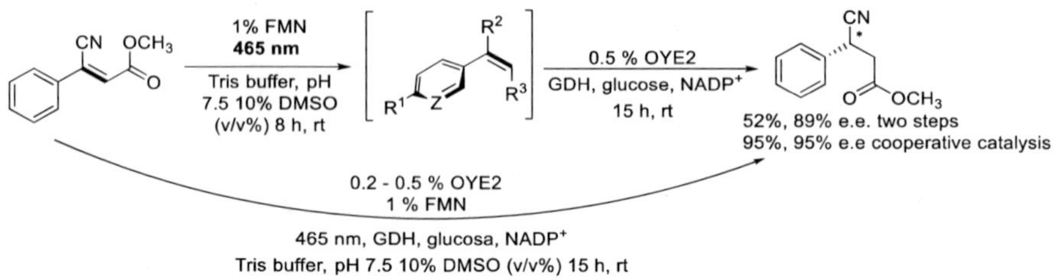

Scheme 11. Sequential vs cooperative photocatalyzed isomerization - enzymatic reduction.

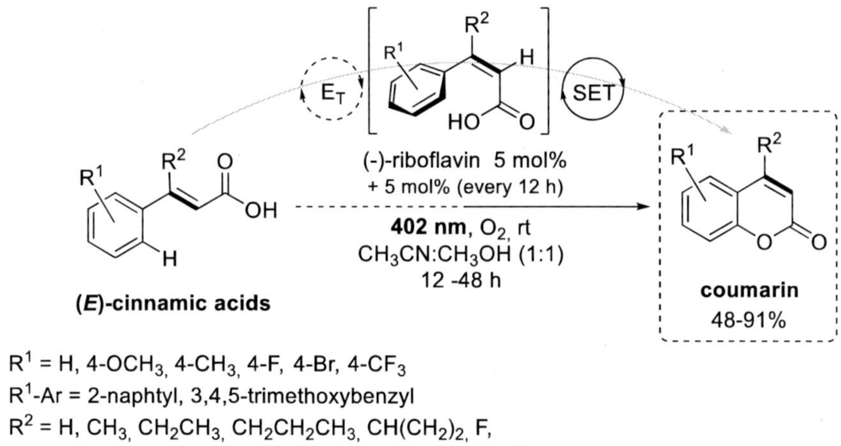

Scheme 12. Single catalyst cascade photocatalysis for the synthesis of coumarins.

The first of this cascade photocatalysis was development by Metternich and Gilmour in 2016 [24.] In their work the authors synthesized coumarins from (*E*)-cinnamic acids through one-photocatalyst-two-activation-modes strategy involving an energy transfer isomerization (E_T) followed by a photoredox cyclization reaction (Scheme 12). The scope of the reaction was good with substitution (EWG and EDG) tolerated in the aromatic ring (48-81%) as well as in the double bond (38-91%).

The proposed mechanism (Scheme 13) was supported by control experiments and reaction progress monitoring through product analysis.

These experiments support the occurrence of a photoisomerization as the first step and the additions of additives (acid and base) support the photoredox cyclization as going through an electron transfer-proton transfer (o viceversa) to the radical intermediate (I) followed by a intramolecular HAS substitution. Kinectic isotopic effect also supported a radical reaction as compared to similar radical intra-HAS "dark" reactions with similar isotopic effect.

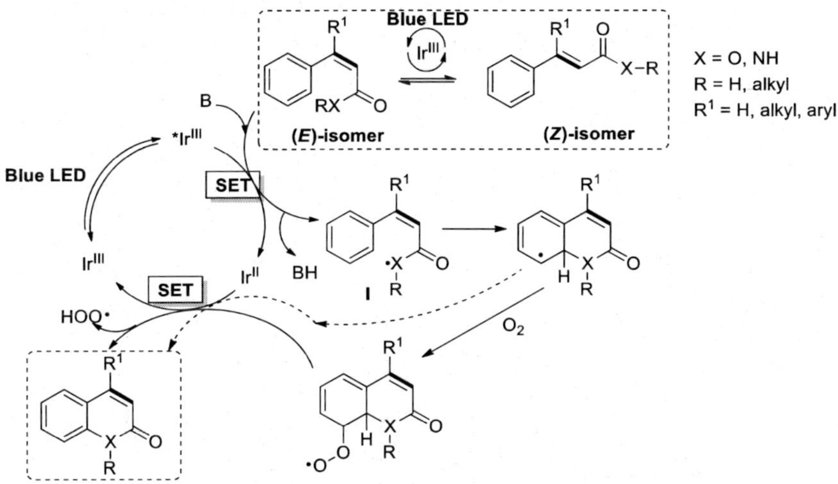

Scheme 13. Proposed mechanism for cascade photocatalysis in cinnamic derivatives.

Analogous reaction with cinnamic amides was demonstrated to be a good synthetic strategy to obtain quinolinones via photoisomerization-photo oxidative C−H amidation [23]. Once again a single catalyst could do the job but, in contrast to cinnamic acid case, a base was necessary to assist the photoredox reaction (Scheme 14). Interestingly, the author optimized the C-H amidation reaction working with *N*-phenylbenzamide to obtain phenanthridinones in which no *E*/*Z* isomerization take place, this initial screening allows the author to proceed to the cinnamic amides and, even when some extra optimization was required, these previous studies shorter the work with the more complex energy transfer-redox system. Photoisomerization was probed by analysis of the *E*/*Z* ratio of cinnamic amides in the absent of any base in which case no C-H amidation occur.

Photoredox catalysis with the occurrence of an amidyl radical was proposed to follow a PCET based in previously reported examples (intermediate I in Scheme 13).

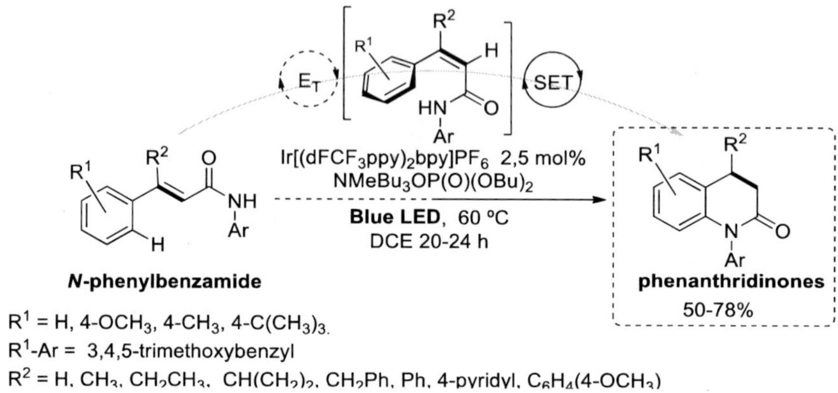

Scheme 14. Single catalyst cascade photocatalysis for the synthesis of quinolinones.

Cycloalkene Photoisomerizations

Direct Excitation

The photochemical property of cycloalkenes is not a simple extension of the acyclic alkenes, particularly with small- and medium-sized rings. In these cases, photoisomerization produce cycloalkenes highly constrained, which are highly reactive and can participate in cycloaddition reactions or nucleophilic additions as well [25].

In a good approach, the photoisomerization of (*Z*)-cycloalkenones, a well-established procedure, could be combined with cycloaddition reaction to access complex polycyclic structures, present in numerous natural products. Datta and co-workers have carried out the synthesis of complex polycyclic structures with (*E*)-fused 6/7 ring system by photoisomerization of (*Z*)-cycloheptenone coupled with facile Diels-Alder cycloaddition (Scheme 15) [26]. In spite of cycloheptenone derivatives are poor dienophiles toward Diels-Alder reactions, the light induced isomerization generate a highly reactive constrained intermediate, which undergoes spontaneous Diels-Alder reaction. Furthermore, if a bicycloheptenone with a defined stereochemistry at the ring fusion is constructed, this could direct cycloheptenone derivatives to isomerize giving rise to two diastereomeric products. With this protocol, an *anti* fusion across the enone moiety of benzofuran ring was observed as main product, allowing access to a family of polycyclic structure with good to excellent isolated yields (60-80%) (Scheme 15).

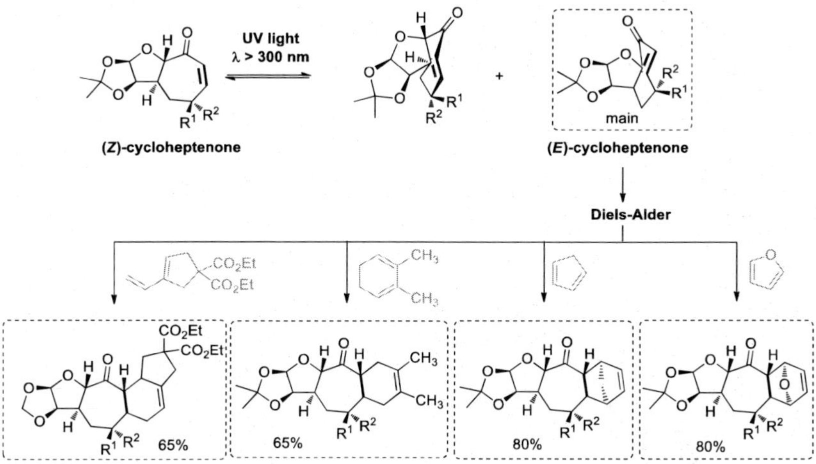

Scheme 15. Diels-Alder cycloaddition of cycloheptenone via photoisomerized activation.

Sensitized Excitation

Considering the highly reactive strained (*E*)-cycloalkene, Beauchemin et al. carried out the intermolecular hydroamination of cycloalkenes with azoles by photosensitized reactions (Scheme 16). The mechanism involves the photoisomerization of (*Z*)-cycloalkene to generate (*E*)-cycloalkene intermediate which quickly undergoes protonation followed by nucleophilic attack of azole. The nature of a this highly strained intermediate allow using catalytic triflic acid as additive in the presence of basic nucleophilic nitrogen of an azole heterocycle. Under this condition, the intermolecular hydroamination of 6- and 7-membered cyclic alkenes occurred in moderate and good yields [27].

For bigger cycloalkenes, such as an 8-member ring, the photoisomerization carries to the formation of *(E)*-cyclooctene which can be isolated under ambient conditions. The high reactivity of *(E)*-cyclooctene allows it to be involved in reactions difficult to access by the *(Z)* isomer. The strained structure of *(E)*-cyclooctene gives it a character of dienophile being able to react with 1,2,4,5-tetrazines (Tz) in an inverse electron demanded Diels–Alder reaction (IEDDA) (Scheme 17) [28,29]. This reaction is exceptionally fast and is frequently used as bioorthogonal ligations, founding numerous applications in drug delivery, fluorescent imaging, radionuclide therapy and several others [30].

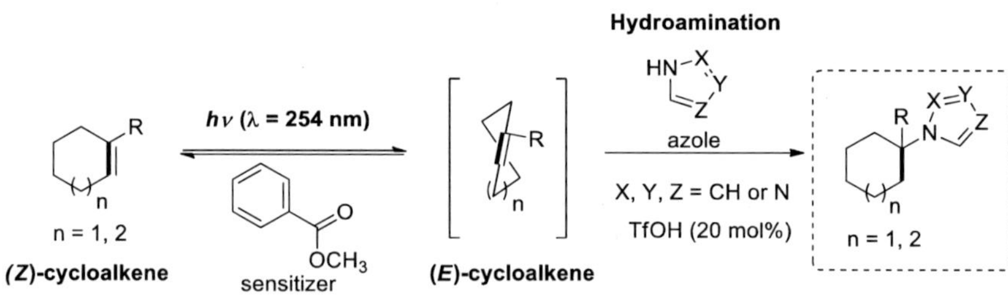

Scheme 16. Nucleophilic attack of azole heterocycles to (*E*)-cycloalkenes.

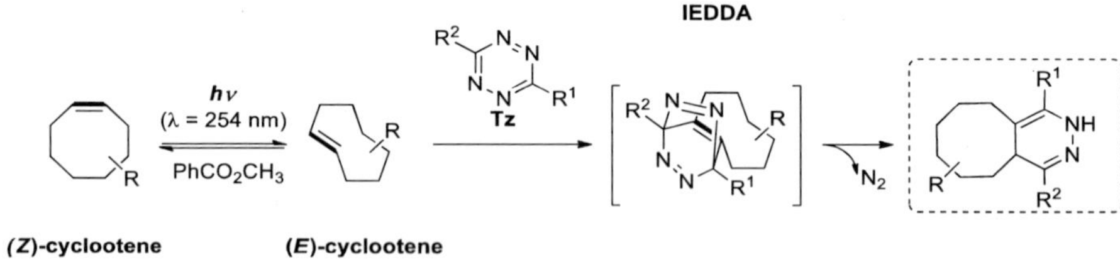

Scheme 17. Inverse electron demand Diels-Alder reaction between (*E*)-cyclooctene and 1,2,4,5-tetrazine.

Due to the importance of *(E)*-cyclooctene as reagent, different derivatives were synthesized depending on the necessities for ligation rate of bioconjugation, stability, solubility, steric impact and availability. Even when there are *(E)*-cyclooctene derivatives commercially available, the increased demand has carried to the development of new synthetic protocols to obtain larger quantities at low cost. In this context, flow photochemistry has emerged as a powerful technique providing the ability to scaling up the reaction, increasing the opportunities for a better control over the conditions, and improving the reaction selectivity and reproducibility. Considering this, Mikula and co-workers built a flow photoreactor of low-cost assembled from readily accessible components. The equipment was evaluated carrying out the photoisomerization of *(Z)*-cyclooctene derivatives frequently used in bioorthogonal chemistry. The isomerization was carried out by singlet sensitization of methyl benzoate under UV irradiation and *(E)*-cyclooctene product was trapped by complexation with silver cations (Scheme 18). The flow-photochemical reaction allows excellent reaction yields in multigram scale in a short time [31]. Recently, Rutjes and co-workers developed a new flow set-up for the synthesis of (*E*)-cyclooctene derivatives which involve the use of liquid-liquid extraction module instead of a silica gel column. This modification involves an improvement in the scale up and speeds up of the photoisomerization, due to the possibility of external addition of a substrate [32].

Furthermore, enantiodifferentiating photochemical transformation sensitized by a chiral catalyst has attracted the interest in the area of asymmetric photochemistry. Particularly, the highly constrained (*E*)-cycloalkenes are known to have optical properties of particular interest and have been the subject of intensive study. Taking this into account, Sugimura and co-workers have been explored the photoisomerization, epoxidation and cyclopropanation of 2,4-pentanediol tethered cyclooctene, to evaluate the stereocontrol ability of the tethered group (Scheme 19) [33]. For the photochemical reaction, the idea is based on the intramolecular sensitization of cyclooctene by the benzoyl moiety through a chiral tether. Their studies showed that the photoisomerization of (2*R*,4*R*)-2,4-pentanediol-tethered cyclooctene achieved a diastereomeric excess (*de*) of 20% at a high *Z*/*E* ratio of 0.80 at 25°C. However, (2*S*,4*R*)-2,4-pentanediol-tethered cyclooctene yielded a lower *de* of 4% in spite of an even higher *Z*/*E* ratio (0.84). These results suggest that the presence of additional photo- and/or thermal processes could reduce the *de* value of the product.

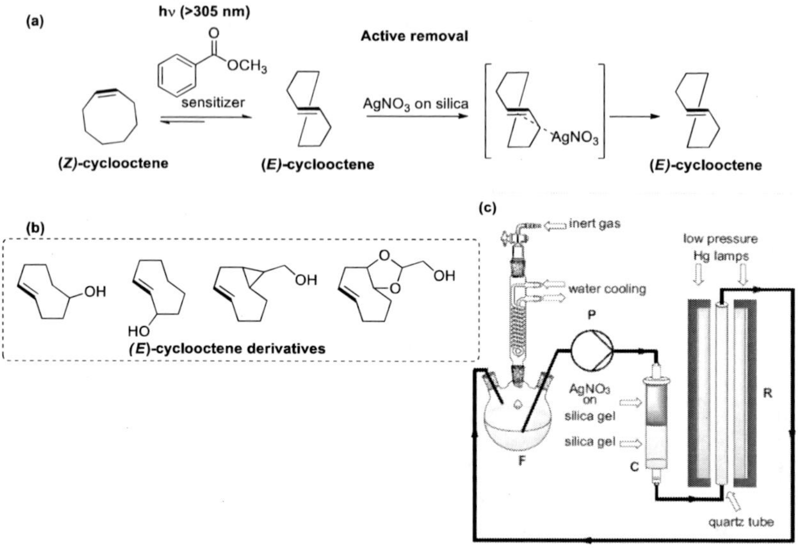

Scheme 18. (a) Photoisomerization of *(Z)*-cyclooctene and complexation of *(E)*-cyclooctene with silver cations. (b) Scope of the reaction under study. (c) Schematic representation of flow-photoreactor used to scale-up the photoisomerization of *(Z)*-cyclooctene.

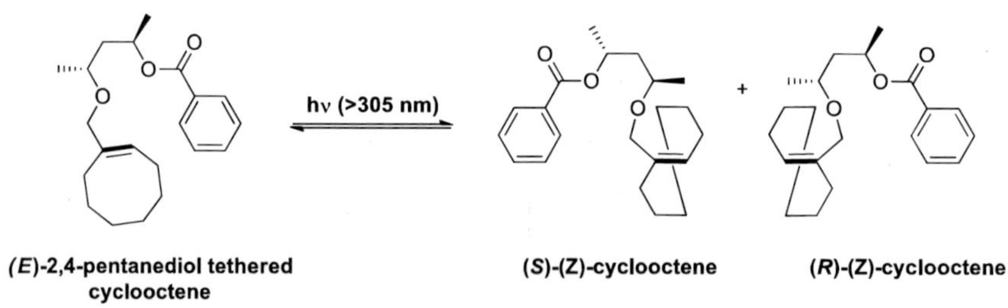

Scheme 19. Enantiodifferentiating photochemical transformation of cyclooctene derivatives.

Although Sugimura's work is clearly a study of stereocontrol evaluation of an intramolecular sensitizated photoisomerization, this kind of researches has made possible the consideration of its application in organic synthesis.

In the literature there are examples of asymmetric light induced isomerization with synthetic utility.

SYNTHESIS OF NATURAL PRODUCTS

As we mention before, photochemical reactions have taken part in many synthetic protocols to obtain molecules with valuable structural complexity in a simple and environmentally mild way. Photoisomerization has received a significant attention as synthetic strategy, and currently we can found interesting examples in natural product synthesis. In this section, we highlight the synthesis of natural product involving a photoisomerization as key step.

In the work from Gilmour and Watson mentioned in section 1.2, the photo-sensitized isomerization of (*E*)-4,4,5,5-tetramethyl-2-(3,4,5-trimethoxystyryl)-1,3,2-dioxaborolane to the (*Z*)-isomer could be used in a stereospecific synthesis of anti-tumor natural product Combretastatin A4 using a Suzuki-Miyaura Pd catalyzed coupling with 4-bromo-2-methoxyphenol (Scheme 20) [18].

Showing the potential of photoisomerization process, the total synthesis of three diastereomers of (-)-cephalimysin A was described in 2013 by Rovis and co-workers, involving a tandem photoisomerization/Stetter reaction (Scheme 21) [34]. Cephalimysin A contains a unique 1-oxa-7-azaspiro[4.4]non-2-ene-4,6-dione core structure and presents a wide range of biological activities [35]. This spirofuranone lactam core can be reached by an intramolecular Stetter reaction with an enal-maleimide derivative (Scheme 21). The olefin geometry of the enal is crucial to the success of the Stetter reaction being the (*Z*)-isomer a necessary intermediate. Since during the enal precursor synthesis the undesired olefin isomer was obtained, a previous alkene isomerization process was required. Therefore, a tandem *in-situ*-photoisomerization/Stetter reaction in the presence of the carbene catalyst was optimized, becoming a key way to achieve the efficient enantioselective synthesis of spirocyclic furanone core. Subsequent conjugate addition, Barbier coupling, oxidation/elimination and oxidation reactions afforded multiple diastereomers of cephalimysin A.

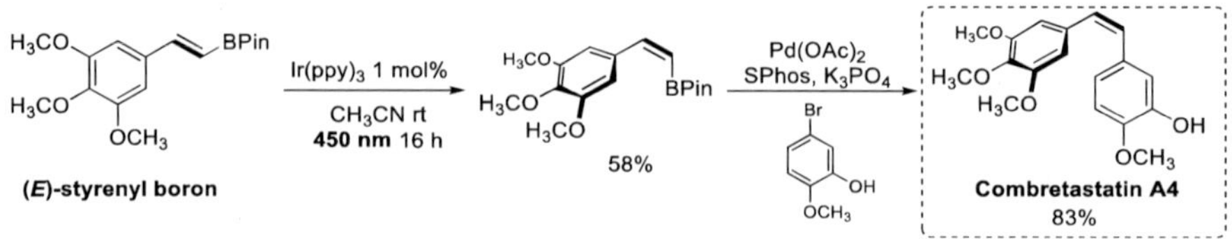

Scheme 20. Synthesis of anti-tumour Combretastatin A4 by photoisomerization, Suzuki-Miyaura Pd catalyzed sequence.

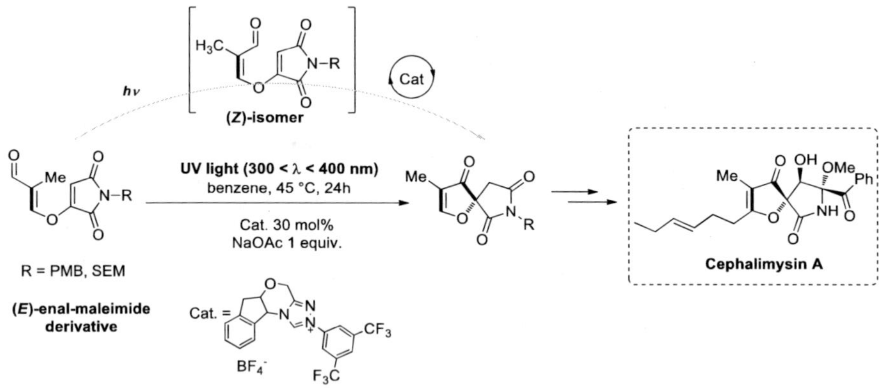

Scheme 21. Synthesis of Cephalimysin A by a tandem photoisomerization/ Stetter reaction.

Another relevant example of a key alkene photoisomerization was developed by Mulzer and co-worker in the total synthesis of 17-deoxyprovidencin (Scheme 22) [36]. The multi-step synthetic sequence includes as key step an aldol addition with subsequent oxidative elimination of selenide (*Z*)-selective ring-closing metathesis (RCM) macrocyclization to obtain product A (scheme 22). The *Z*/*E* photoisomerization of A to a highly strained conformationally restricted ring system allowed the stereoselective epoxidation of the *E* olefin to obtain the right 17-deoxyprovidencin product. It should be noted that in order to succeed in the photoisomerization stage the use of Pyrex glass was crucial, allowing selective excitation of the (*Z*)-isomer (cuts off below 300 nm). This target molecule is a direct precursor of providencin, a class of furanocembranoid natural product which exhibits moderate activity against human breast cancer and lung and central nervous system cancer cell lines [37].

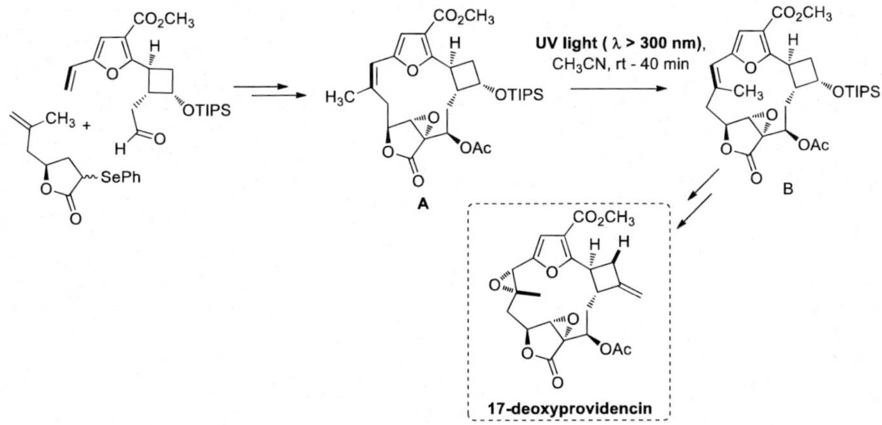

Scheme 22. Synthesis of 17-Deoxyprovidencin.

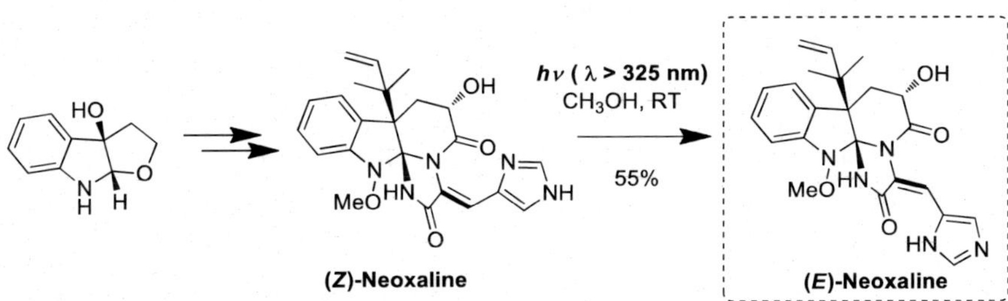

Scheme 23. Synthesis and photoisomerization of unnatural (*Z*)-neoxaline to natural product (*E*)-neoxaline.

Alkene photoisomerization reactions have also been used as a late stage in obtaining the right stereoisomer of a target natural product. The mild reaction conditions employed do not affect other functionalities present in the molecule making them attractive for this purpose. Following this approach, Sunazuka and co-workers developed the first asymmetric total synthesis of (*E*)-neoxaline [38], a biologically active prenylated indole alkaloid possessing a unique indoline spiroaminal framework. The strategy involves the generation of (*Z*)-neoxaline isomer through a sequence of reactions including radical allylation, stepwise oxidations and cyclizations from indoline derivative, and aldol reaction followed by elimination (Scheme 23). Hence, photoisomerization of unnatural (*Z*)-neoxaline to natural (*E*)-neoxaline was needed as the crucial last step. A short-pass filter (λ > 325 nm) was used to perform selective excitation of the unnatural *Z*-isomer, this avoid the formation of unfavorable equilibrium mixtures improving the efficiency of this step affording natural (*E*)-neoxaline in good yield (55%) (Scheme 23).

In the same sense, Harran and co-workers carried out the synthesis of (-)-Callyspongiolide involving a sequence of 18 steps, which presents two strategic steps of a chiral iridium catalysis and a photoisomerization. In spite of the large synthetic route, the design involves the build of three key segments with high yields, with the possibility to create easily structure diversity on callyspongiolide isomers. Specifically, the photoisomerization of (*E*,*E*)-hydroxy urethane was carried out under sensitized condition, irradiated at 300 nm. After 20 minute of irradiation, the photoequilibrium was established, being able to access to desired (*Z*)-isomer in 55% isolated yield (Scheme 24). In order to obtain large amounts of (-)-Callyspongiolide, the recovered (*E*)-isomer was resubjected to photoisomerization [39].

The photoisomerization of cycloalkenes has been successfully applied to the formation of natural product, proving the scope of this photochemical reaction. One clear example of these was recently informed by Tang and co-workers, who have achieved the synthesis of 4*β*,5*β*-epoxyxanthatin-1*α*,4*α*-endoperoxide (a novel xanthanolide) by sequential double-bond isomerization, 6π electronic cyclization, and singlet-oxygen mediated [4+2] cycloaddition from xanthatin (Scheme 25) [40]. Xanthanolides are compounds with impressive biological activities, such as anticarcinogenic, antimycotic, and antibacterial, which makes them an important target for natural product synthesis [41]. Since xanthatin has two different alkene moieties, it was crucial for this transformation to obtain regioselectively the C1–C5 instead of C2–C3 double-bond isomerization. The C1–C5 double-bond isomerization was favored by the use of DBU during the photochemical induction, which promotes 1,6-conjugate addition/δ-elimination on the (E)-stereoisomer and generates the right intermediate for the tandem 6π electronic cyclization.

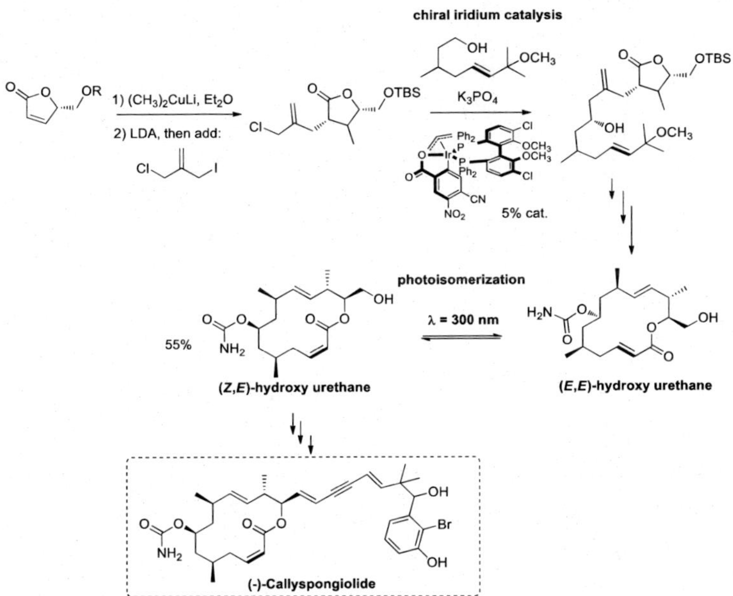

Scheme 24. Total synthesis of (-)-Callyspongiolide natural product.

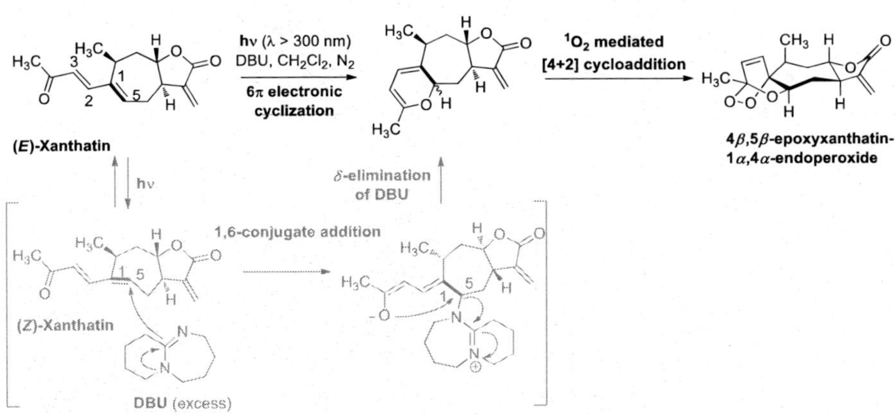

Scheme 25. Synthesis of 4β,5β-epoxyxanthatin-1α,4α-endoperoxide.

The authors synthesized four new mogolide derivatives, products of dimerization reaction of (*Z*)-xanthatin, which include two authentic natural products (mogolide A and B). This works is a clear example of a biomimetic synthesis guided toward natural product discovery by photochemical transformations. The proposed mechanism involve as the key step the photoisomerization reaction followed by [4+2] cycloaddition to obtain mogolide B and mogolide C compounds, or [4+2]/[2+2] cycloaddition to generate mogolide A and 3-epi-mogolide A (Scheme 26).

In the last years, flow photochemistry has been successfully applied to natural products or molecule target synthesis, as a safer and more sustainable synthetic protocol having a high impact on the pharmaceutical industry and medicine [42]. For this reason, it is not surprising that alkene photoisomerization applied to natural product synthesis have also been studied under continuous-flow condition.

In this way, Fox and co-workers have performed the total synthesis of hyacinthacine A2 using a flow photoreactor [43], a pyrrolizidine alkaloid with biological activity against HIV [44] (Scheme 27). The access to the pyrrolizidine scaffold was afforded through transannular hydroaminations of 5-aza-(*E*)-cyclooctene. The *E* stereoisomer arises from the photoisomerization of 5-aza-(*Z*)-cyclooctene derivative and the key step to obtain the target natural product is the stereocontrol in this photoisomerization step. High diastereoselectivity was observed in the photoisomerization when an acetonide derivative of 5-aza-(*Z*)-cyclooctene was employed, due to conformational constraints of the ring fusion. This flow system with a very unfavorable *Z*/*E* ratio is driven to completion by active removal of the (*E*)-isomers through a stable complexes with $AgNO_3$. The flow set-up used is similar that showed in the Scheme 18. The subsequent easy acetonide removal by acid treatment, allows access to the right (*E*)-stereoisomer required for the transannulation reaction. Finally, the absolute planar chirality of 5-aza-(*E*)-cyclooctene is transferred to point chirality in the product with excellent fidelity: only a single diastereomer of hyacinthacine A2 is obtained in the transannular reaction.

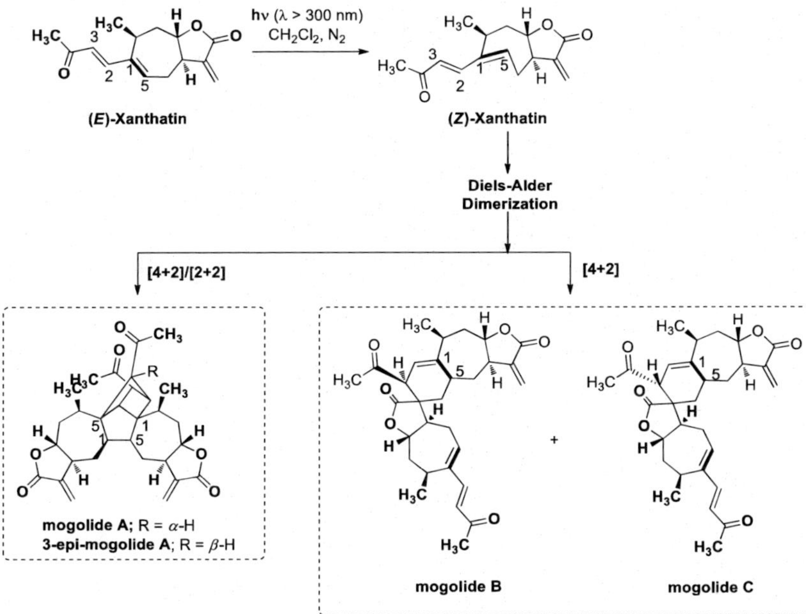

Scheme 26. Dimerization of xanthatin derivative through photoisomerization/Diels-Alder sequence.

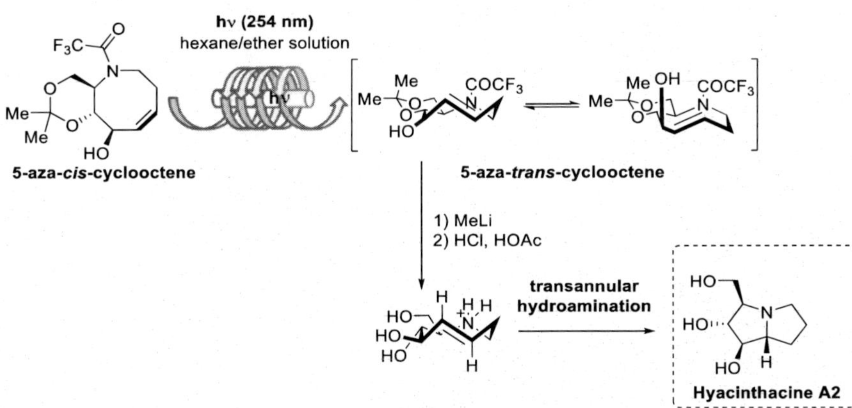

Scheme 27. Synthesis of hyacinthacine A2 alkaloid by transannulation reaction in flow.

Cantrell and co-workers has developed the synthesis of doxercalciferol from commercially available ergocalciferol. The authors used a continuous photoisomerization of a known vitamin D intermediate as the key step [45].

After five step of reaction, without purification of the intermediates, ergocalciferol was modified to the key intermediate (*E*,*E*)-triene which crystallize due to the presence of appropriate protecting groups. The continuous photoisomerization of the (*E*,*E*)-triene with 9-acetylanthracene sensitizer, allowed them to obtain the triene with the desired geometry in 95% yield. They also could couple the removal of the sensitizer from the product solution with an in-line carbon/Celite column placed after the photoreactor. Treatment of crude product with HCl in the presence of alcohol solvent, cleanly afforded crude doxercalciferol. Further purification of the crude product by crystallization afforded high-purity doxercalciferol from ergocalciferol in about 10% overall yield without the need for chromatography.

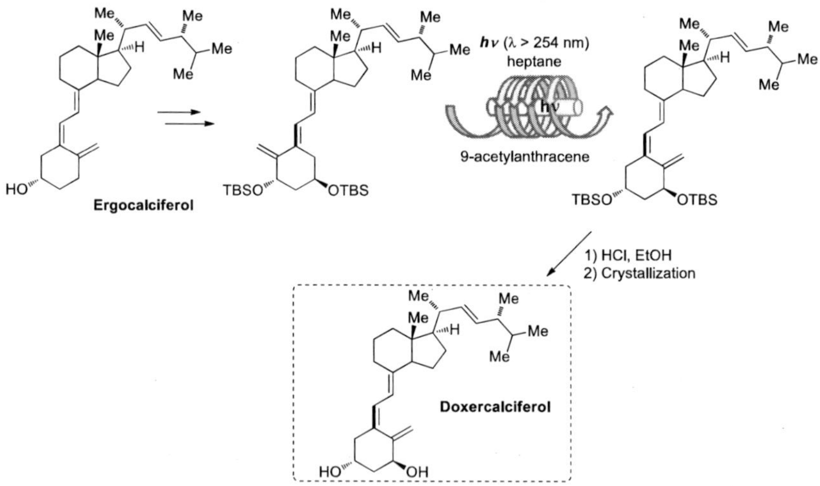

Scheme 28. Photoisomerization in flow for the synthesis of doxercalciferol.

Conclusion

Photochemical mediated isomerizations are a useful tool for construction of complex molecular frameworks, many times in a simple way. Besides direct irradiation as well as sensitized reactions are useful for synthetic purposes with UV and visible light sources. For acyclic compounds the double bonds isomerization is mainly applied to access to a specific alkene geometry which is directly involved in the desired transformation. On the other hand, in cycloalkenes the photoisomerization usually produce a highly constrained and more reactive intermediate, allowing reactions that otherwise will be challenging to access.

The versatility of the photoisomerization reactions remain reflected in its use as a key step of total synthesis of natural products and biologically active compounds. Additionally, with the technological advances, the use of flow photochemistry has allowed to carry out light activated isomerization with high yields and selectivity, in safer and more sustainable conditions.

Cooperative cascade catalytic transformations deserve a special mention since they constitute a simple and efficient way to perform complex transformation in high yields. The importance of these strategies is evident in partially and unfavorable *Z*/*E* isomerization that could be find in many transformation.

Thus, in combination with multi-step processes, photoisomerization will continue being an attractive synthetic tool to access to complex molecular structures with biological and medicinal importance.

REFERENCES

[1] Kärkäs, M. D.; Porco, J. A.; Stephenson, C. R. J. Photochemical Approaches to Complex Chemotypes: Applications in Natural Product Synthesis. *Chem. Rev.* 2016, *116* (17), 9683–9747.

[2] Staveness, D.; Bosque, I.; Stephenson, C. R. J. Free Radical Chemistry Enabled by Visible Light-Induced Electron Transfer. *Acc. Chem. Res.* 2016, *49* (10), 2295–2306.

[3] *Handbook of Synthetic Photochemistry;* Albini, A., Fagnoni, M., Eds.; Wiley-VCH Verlag GmbH & Co. KGaA: Weinheim, Germany, 2009.

[4] Kaneko, A.; Inoue, K.; Kojima, K.; Kandori, H.; Sudo, Y. Conversion of Microbial Rhodopsins: Insights into Functionally Essential

Elements and Rational Protein Engineering. *Biophys. Rev.* 2017, *9* (6), 861–876.

[5] Stefanachi, A.; Leonetti, F.; Pisani, L.; Catto, M.; Carotti, A. Coumarin: A Natural, Privileged and Versatile Scaffold for Bioactive Compounds. *Molecules* 2018, *23* (2), 250.

[6] *Synthetic Organic Photochemistry;* Griesbeck, A. G., Mattay, J., Eds.; Marcel Dekker: New York, U.S.A., 2005.

[7] Thomé, I.; Besson, C.; Kleine, T.; Bolm, C. Base-Catalyzed Synthesis of Substituted Indazoles under Mild, Transition-Metal-Free Conditions. *Angew. Chemie Int. Ed.* 2013, *52* (29), 7509–7513.

[8] Li, B.; Williams, B. D.; Smith, A. B. A Mild, Diastereoselective Construction of Cyclic and Spirocyclic Ketals Employing a Tandem Photoisomerization/Cyclization Tactic. *Org. Lett.* 2015, *17* (1), 3–5.

[9] Chen, X.; Qiu, S.; Wang, S.; Wang, H.; Zhai, H. Blue-Light-Promoted Carbon-Carbon Double Bond Isomerization and Its Application in the Syntheses of Quinolines. *Org. Biomol. Chem.* 2017, *15* (30), 6349–6352.

[10] Sharma, V.; Gupta, V.; Anthal, S.; Saxena, A. K.; Ishar, M. P. S. Photochemical Formation and Decomposition of 8-[β-Arylethenyl]-2,2,6-Trimethyl-7,9,10-Trioxa-Tricyclo[6.2.2.01,6]Dodec-11-Ene to Novel 6-Hydroxy-1,7,7-Trimethyl-2-Oxa-Bicyclo[4.4.0]Dec-4-En-3-One in the Presence of Oxygen. *Tetrahedron Lett.* 2012, *53* (42), 5649–5651.

[11] Gandhi, R. P.; Aryan, R. C. Photochemical Reorganisation of 1,7,7-Trimethyl-3-(E-2′-Arylethenyl)-2-Oxabicyclo[4.4.0]Deca-3,5-Dienes. *J. Chem. Soc., Chem. Commun.* 1988, No. 15, 1024–1025.

[12] Gandhi, R. P.; Kumar, S.; Aryan, R. C.; Ishar, M. P. S. Regioselective Photoelectrocyclization of (E/E)-Arylidene-β-Ionones; Synthesis of Some 1,7,7-Trimethyl-3-(e-2′-Arylethenyl)-2-Oxabicyclo [4.4.0] Deca-3,5-Dienes. *Synth. Commun.* 1989, *19* (9–10), 1759–1762.

[13] Ishar, M. P. S.; Singh, R.; Kumar, K.; Singh, G.; Velmurugan, D.; Pandi, A. S.; Raj, S. S. S.; Fun, H. K. Photochemistry of Arylidene- β -Ionones: A Highly Efficient Route to Novel Tricyclic Ketones through Intramolecular, Exoselective Photochemical (4 + 2)

Cycloadditions, Occurring Only in an Aqueous−Organic Solvent. *J. Org. Chem.* 2002, *67* (7), 2234–2240.

[14] Zhan, K.; Li, Y. Visible-Light Photocatalytic E to Z Isomerization of Activated Olefins and Its Application for the Syntheses of Coumarins. *Catalysts* 2017, *7* (11), 337.

[15] Horaguchi, T.; Hosokawa, N.; Tanemura, K.; Suzuki, T. Photocyclization Reactions. Part 8. Synthesis of 2-Quinolone, Quinoline and Coumarin Derivatives Using Trans-Cis Isomerization by Photoreaction. *J. Heterocycl. Chem.* 2002, *39* (1), 61–67.

[16] Faßbender, S. I.; Metternich, J. B.; Gilmour, R. Spatiotemporal Control of Pre-Existing Alkene Geometry: A Bio-Inspired Route to 4-Trifluoromethyl-2 H -Chromenes. *Org. Lett.* 2018, *20* (3), 724–727.

[17] Arcadi, A.; Cacchi, S.; Fabrizi, G.; Marinelli, F.; Verdecchia, M. 4-Aryl- and 4-Vinyl-2,2-Dialkyl-3-Chromenes from Tertiary 3-(o -Bromophenyl)Propynols via a Palladium-Catalyzed Hydroarylation/ Hydrovinylation-Cyclization Sequence. *Synlett* 2006, 909–915.

[18] Molloy, J. J.; Metternich, J. B.; Daniliuc, C. G.; Watson, A. J. B.; Gilmour, R. Contra-Thermodynamic, Photocatalytic E → Z Isomerization of Styrenyl Boron Species: Vectors to Facilitate Exploration of Two-Dimensional Chemical Space. *Angew. Chemie* 2018, *130* (12), 3222–3226.

[19] Litman, Z. C.; Wang, Y.; Zhao, H.; Hartwig, J. F. Cooperative Asymmetric Reactions Combining Photocatalysis and Enzymatic Catalysis. *Nature* 2018, *560* (7718), 355–359.

[20] Zhao, F.; Li, N.; Zhang, T.; Han, Z.-Y.; Luo, S.-W.; Gong, L.-Z. Enantioselective Aza-Ene-Type Reactions of Enamides with Gold Carbenes Generated from α-Diazoesters. *Angew. Chemie Int. Ed.* 2017, *56* (12), 3247–3251.

[21] Kong, D.; Li, M.; Wang, R.; Zi, G.; Hou, G. Highly Efficient Asymmetric Hydrogenation of Cyano-Substituted Acrylate Esters for Synthesis of Chiral γ-Lactams and Amino Acids. *Org. Biomol. Chem.* 2016, *14* (4), 1216–1220.

[22] Chung, Yu-Chiang; Janmanchi, D.; Wu, H. Preparation of Chiral 3-Arylpyrrolidines via the Enantioselective 1,4-Addition of Arylboronic

Acids to Fumaric Esters Catalyzed by Rh(I)/Chiral Diene Complexes. *Org. Lett.* 2012, *14* (11), 2766–2769.

[23] Moon, Y.; Jang, E.; Choi, S.; Hong, S. Visible-Light-Photocatalyzed Synthesis of Phenanthridinones and Quinolinones via Direct Oxidative C-H Amidation. *Org. Lett.* 2018, *20* (1), 240–243.

[24] Metternich, J. B.; Gilmour, R. One Photocatalyst, n Activation Modes Strategy for Cascade Catalysis: Emulating Coumarin Biosynthesis with (−)-Riboflavin. *J. Am. Chem. Soc.* 2016, *138* (3), 1040–1045.

[25] *CRC Handbook of Organic Photochemistry and Photobiology. 2da Edition*; William Horspool and Francesco Lenci, Ed.; Washington DC, USA, 2004.

[26] Ghosh, S.; Bose, S.; Jana, A.; Nijamudheen, A.; Datta, A. Influence of Ring Fusion Stereochemistry on the Stereochemical Outcome in Photo-Induced Diels–Alder Reaction of Fused Bicycloheptenone Derivatives. *Tetrahedron* 2014, *70* (52), 9783–9790.

[27] Moran, J.; Cebrowski, P. H.; Beauchemin, A. M. Intermolecular Hydroaminations via Strained (E)-Cycloalkenes. *J. Org. Chem.* 2008, *73* (3), 1004–1007.

[28] Royzen, M.; Yap, G. P. A.; Fox, J. M. A Photochemical Synthesis of Functionalized Trans -Cyclooctenes Driven by Metal Complexation. *J. Am. Chem. Soc.* 2008, *130* (12), 3760–3761.

[29] Blackman, M. L.; Royzen, M.; Fox, J. M. Tetrazine Ligation: Fast Bioconjugation Based on Inverse-Electron-Demand Diels−Alder Reactivity. *J. Am. Chem. Soc.* 2008, *130* (41), 13518–13519.

[30] Blackman, M. L.; Royzen, M.; Fox, J. M. Tetrazine Ligation: Fast Bioconjugation Based on Inverse-Electron-Demand Diels−Alder Reactivity. *J. Am. Chem. Soc.* 2008, *130* (41), 13518–13519.

[31] Svatunek, D.; Denk, C.; Rosecker, V.; Sohr, B.; Hametner, C.; Allmaier, G.; Fröhlich, J.; Mikula, H. Efficient Low-Cost Preparation of Trans-Cyclooctenes Using a Simplified Flow Setup for Photoisomerization. *Monatshefte für Chemie - Chem. Mon.* 2016, *147* (3), 579–585.

[32] Blanco-Ania, D.; Maartense, L.; Rutjes, F. P. J. T. Rapid Production of Trans -Cyclooctenes in Continuous Flow. *Chem Photo Chem* 2018, *2* (10), 898–905.

[33] Song, K.; Fujita, M.; Okuyama, T.; Sugimura, T. Stereoselective E/Z-Photoisomerization of Cyclooctene Using 2,4-Pentanediol Chiral Tether. *Tetrahedron: Asymmetry* 2017, *28* (2), 296–301.

[34] Lathrop, S. P.; Rovis, T. A Photoisomerization-Coupled Asymmetric Stetter Reaction: Application to the Total Synthesis of Three Diastereomers of (−)-Cephalimysin A. *Chem. Sci.* 2013, *4* (4), 1668–1673.

[35] Yamada, T.; Imai, E.; Nakatuji, K.; Numata, A.; Tanaka, R. Cephalimysin A, a Potent Cytotoxic Metabolite from an Aspergillus Species Separated from a Marine Fish. *Tetrahedron Lett.* 2007, *48* (36), 6294–6296.

[36] Toelle, N.; Weinstabl, H.; Gaich, T.; Mulzer, J. Light-Mediated Total Synthesis of 17-Deoxyprovidencin. *Angew. Chemie - Int. Ed.* 2014, *53* (15), 3859–3862.

[37] Marrero, J.; Rodríguez, A. D.; Baran, P.; Raptis, R. G. Isolation and Structure of Providencin: A Highly Oxygenated Diterpene Possessing a Unique Bicyclo[12.2.0]Hexadecane Ring System from the Sea Plume Pseudopterogorgia Kallos. *Org. Lett.* 2003, *5* (14), 2551–2554.

[38] Ideguchi, T.; Yamada, T.; Shirahata, T.; Hirose, T.; Sugawara, A.; Kobayashi, Y.; Ōmura, S.; Sunazuka, T. Asymmetric Total Synthesis of Neoxaline. *J. Am. Chem. Soc.* 2013, *135* (135), 12568–12571.

[39] Manoni, F.; Rumo, C.; Li, L.; Harran, P. G. Unconventional Fragment Usage Enables a Concise Total Synthesis of (-)-Callyspongiolide. *J. Am. Chem. Soc.* 2018, *140* (4), 1280–1284.

[40] Shang, H.; Liu, J.; Bao, R.; Cao, Y.; Zhao, K.; Xiao, C.; Zhou, B.; Hu, L.; Tang, Y. Biomimetic Synthesis: Discovery of Xanthanolide Dimers. *Angew. Chemie Int. Ed.* 2014, *53* (52), 14494–14498.

[41] Vasas, A.; Hohmann, J. Xanthane Sesquiterpenoids: Structure, Synthesis and Biological Activity. *Nat. Prod. Rep.* 2011, *28* (4), 824.

[42] Politano, F.; Oksdath-Mansilla, G. Light on the Horizon: Current Research and Future Perspectives in Flow Photochemistry. *Org. Process Res. Dev.* 2018, *22* (9), 1045–1062.

[43] Royzen, M.; Taylor, M. T.; DeAngelis, A.; Fox, J. M. Total Synthesis of Hyacinthacine A2: Stereocontrolled 5-Aza-Cyclooctene Photoisomerization and Transannular Hydroamination with Planar-to-Point Chirality Transfer. *Chem. Sci.* 2011, *2* (11), 2162.

[44] Taylor, D. L.; Nash, R.; Fellows, L. E.; Kang, M. S.; Tyms, A. S. Naturally Occurring Pyrrolizidines: Inhibition of α-Glucosidase 1 and Anti-HIV Activity of One Stereoisomer. *Antivir. Chem. Chemother.* 1992, *3* (5), 273–277.

[45] Anderson, B. G.; Bauta, W. E.; Cantrell, W. R. Development of an Improved Process for Doxercalciferol via a Continuous Photochemical Reaction. *Org. Process Res. Dev.* 2012, *16* (5), 967–975.

ABOUT THE EDITOR

Diego Sampedro, PhD
Associate Professor
Universidad de La Rioja, Spain

Diego Sampedro received his PhD degree (2001) from the Universidad de La Rioja for a work focused in the synthesis and reactivity of organometallic species which yielded five and three-membered heterocycles. This study combined synthesis, mechanistic and photochemical studies among other techniques including some theoretical calculations. Synthetic photochemistry was also at the spotlight in the discovery of a brand new photoreaction that involved the rearrangement of N-cyclopropylimines to pyrrolines. After a postdoctoral stay in the Università degli Studi di Siena working in computational organic photochemistry, he returned to Spain as Assistant Professor. In 2004, he obtained a "Ramón y Cajal" research contract in the Universidad de La Rioja and since 2009 he is Associate Professor. His research focuses on organic photochemistry both theoretical and experimental including the design of molecular photoswitches, photoprotective compounds and biological applications of photoactive species. In addition, he is active in the field of popularization of science.

INDEX

#

A

B

C

D

E

F

G

H

I

L

M

N

O

P

Q

R

S

T

U

V

W

Related Nova Publications

Chemical Elements (Fluorine, Rhodium and Rubidium): Properties, Synthesis and Applications

Editor: Adrian Huff

Series: Chemistry Research and Applications

Book Description: Fluorination is one of the most efficient routes for modifying the physiochemical properties of materials in order to enlarge their application domain. In this collection, controlled fluorination was successfully applied for the fluorination of different starting materials: SWCNTs, MWCNTs, carbon nanodiscs, carbon blacks, polymers, carbides, and Li4Ti5O12 (electrochemical materials).

Softcover ISBN: 978-1-53614-017-0
Retail Price: $95

Solution Chemistry: Advances in Research and Applications

Editor: Yongliang Xiong, Ph.D.

Series: Chemistry Research and Applications

Book Description: The aim of this book is to present original data and comprehensive reviews of state-of-the-art work that advance our understanding of solution chemistry in diverse fields, authored by the diverse experts.

Hardcover ISBN: 978-1-53613-101-7
Retail Price: $160